T0136872

Fungal Biology

Series Editors

Dr. Vijai Kumar Gupta
Biorefining and Advanced Materials Research Center
Scotland's Rural College (SRUC), SRUC Barony Campus, Parkgate
Dumfries, Scotland,
United Kingdom

Maria G. Tuohy
School of Natural Sciences
National University of Ireland Galway
Galway, Ireland

About the Series

Fungal biology has an integral role to play in the development of the biotechnology and biomedical sectors. It has become a subject of increasing importance as new fungi and their associated biomolecules are identified. The interaction between fungi and their environment is central to many natural processes that occur in the biosphere. The hosts and habitats of these eukaryotic microorganisms are very diverse; fungi are present in every ecosystem on Earth. The fungal kingdom is equally diverse, consisting of seven different known phyla. Yet detailed knowledge is limited to relatively few species. The relationship between fungi and humans has been characterized by the juxtaposed viewpoints of fungi as infectious agents of much dread and their exploitation as highly versatile systems for a range of economically important biotechnological applications. Understanding the biology of different fungi in diverse ecosystems as well as their interactions with living and non-living is essential to underpin effective and innovative technological developments. This series will provide a detailed compendium of methods and information used to investigate different aspects of mycology, including fungal biology and biochemistry, genetics, phylogenetics, genomics, proteomics, molecular enzymology, and biotechnological applications in a manner that reflects the many recent developments of relevance to researchers and scientists investigating the Kingdom Fungi. Rapid screening techniques based on screening specific regions in the DNA of fungi have been used in species comparison and identification, and are now being extended across fungal phyla. The majorities of fungi are multicellular eukaryotic systems and therefore may be excellent model systems by which to answer fundamental biological questions. A greater understanding of the cell biology of these versatile eukaryotes will underpin efforts to engineer certain fungal species to provide novel cell factories for production of proteins for pharmaceutical applications. Renewed interest in all aspects of the biology and biotechnology of fungi may also enable the development of "one pot" microbial cell factories to meet consumer energy needs in the 21st century. To realize this potential and to truly understand the diversity and biology of these eukaryotes, continued development of scientific tools and techniques is essential. As a professional reference, this series will be very helpful to all people who work with fungi and should be useful both to academic institutions and research teams, as well as to teachers, and graduate and postgraduate students with its information on the continuous developments in fungal biology with the publication of each volume.

More information about this series at http://www.springer.com/series/11224

Xiaofeng Dai • Minaxi Sharma • Jieyin Chen
Editors

Fungi in Sustainable Food Production

Springer

Editors
Xiaofeng Dai
State Key Laboratory for Biology
of Plant Diseases and Insect Pests Institute
of Plant Protection
Chinese Academy of Agricultural Sciences
Beijing, China

Minaxi Sharma
Department of Food Technology
Akal College of Agriculture Eternal
University
Baru Sahib, Himachal Pradesh, India

Jieyin Chen
State Key Laboratory for Biology
of Plant Diseases and Insect Pests Institute
of Plant Protection
Chinese Academy of Agricultural Sciences
Beijing, China

ISSN 2198-7777   ISSN 2198-7785   (electronic)
Fungal Biology
ISBN 978-3-030-64408-6   ISBN 978-3-030-64406-2   (eBook)
https://doi.org/10.1007/978-3-030-64406-2

This Springer imprint is published by the registered company Springer Nature Switzerland AG
The registered company address is: Gewerbestrasse 11, 6330 Cham, Switzerland

Foreword

Food production, storage, transportation, and processing will all change according to the same forces that are affecting other facets of the economy. All will benefit from studies of fungal biotechnology designed for enhancement of food production, security, and safety.

Fungal biotechnology harnesses the potential of fungi while ensuring that our food remains safe, free from fungal pathogens, mycotoxins, mold allergens, and other problems associated with quality loss during storage. Fungal biotechnology has the potential of being a primary driver of global quality food production.

The concepts and practical applications of fungal biotechnology continue to make significant contributions to food science and technology. Without exception, many aspects of food ingredient and process technologies involving fungi are impacted by research in biotechnology; these include functional foods, nutraceuticals, value-added foods, food pathogens, and safety detection systems.

In this manner, the book *Fungi in Sustainable Food Production* was expertly edited by Xiaofeng Dai, Minaxi Sharma, and Jieyin Chen. The volume represents an important contribution to food science and technology advances using fungi.

The 12 chapters of this book work present relevant information that can be useful to students, teachers, researchers, and professionals interested in uses of fungi in food production.

Laboratory of Microbiology and Bioprocesses   Helen Treichel
Federal University of Fronteira Sul
Erechim, RS, Brazil

Preface

This book deals with the introduction and importance of the fungal world in sustainable food production by involving improvements in fungal biology and biotechnology. Fungi are one of the most valuable candidates for the biodiscovery of novel food ingredients, additives, enzymes, drugs, and antimicrobials, and play an important role in environmental sustainability. Nowadays, fungi have been pioneered in the food sector for manufacturing many bioactive components of food importance and have become a routine part of human life. Metabolic versatility to produce various kinds of food products makes fungi interesting objects for the range of economically important biomolecules for food biotechnology and related applications. In addition, fungal biotechnology promotes food quality and safety. With this book, we aim to include insights in the use of fungi for food processing of biobased products. This book offers a comprehensive reference guide to the field of fungal-food production and elicits important facets such as technological aspects and challenges during production of food-bio-nutrients like dietary fiber, proteins, enzymes, polyunsaturated fatty acids, single cell proteins, and other fermented products. The 12 chapters cover various fungal-food nutrients, their production, characterization, and industrial importance. The primary objective of this book is to compile a valuable resource on the challenges and potential of fungal contribution in the agri-food sector. However, there is a gap in this area, especially on topics related to prospects and investment, as well as intellectual and technical issues. Therefore, this publication is a bridge that links the gaps compared to the already-available ones on the role of fungal systems in food sustainability.

The book is intended for all researchers, academics, and students who are interested in the current trends and future prospects of sustainable food production through fungi and fungal bioprocesses. We are thankful to authors of all the chapters for their efficient cooperation and readiness in revising the manuscripts. We would also like to extend our appreciation to the reviewers who in spite of their busy schedules assisted us by evaluating the manuscripts and provided their critical comments to improve them. We sincerely thank Dr. Vijai Kumar Gupta and Dr. Maria G. Tuohy and the team at Springer Nature for their cooperation and efforts in producing this book.

Beijing, China   Xiaofeng Dai
Baru Sahib, Himachal Pradesh, India   Minaxi Sharma
Beijing, China   Jieyin Chen

Contents

About the Editors

Xiaofeng Dai is currently working as the Group Leader of the Crop Fungi Disease in Institute of Plant Protection and The Director General of Feed Research Institute, CAAS (Chinese Academy of Agricultural Sciences), and took charge The Director General of Institute of Food Science and Technology, CAAS (2010 October–2019 May). He supervised Doctoral students, has made outstanding contributions: national agricultural scientific research talents, extinguish young expert appointed by the Ministry of Agriculture of The People's Republic of China, recipient of the special allowance from the State Council, chief scientist of team of the national agricultural science and technology innovation project, leading talent of the Academy of Agricultural Sciences and many more. Dr. Dai had presided over more than 20 national issues, won 4 of National Science and Technology Awards, 6 Provincial Science and Technology Awards, the National "Five-year" Science and Technology Major Achievements Award, "National Five-year Science and Technology Scientist" Award from the state council and contribution awards of "national medium and long-term planning of science and technology strategy", and published more than 200 papers. His research team had been awarded as "outstanding scientific research innovation team of Ministry of Agriculture".

Dr. Minaxi Sharma is currently working as ERA Chair VALORTECH-Senior Research Scientist, Estonian University of Life Sciences (EMU), Tartu, Estonia. Before joining EMU- Estonia, she worked as Assistant Professor for two years with teaching and research responsibilities, in the Department of Food Technology, Eternal University, Baru Sahib, Himachal Pradesh, India. She obtained her doctorate degree in Dairy Chemistry from National Dairy Research Institute, Karnal (India). She has 9+ years of extensive experience in the research area of: applications of antimicrobial nanoencapsulation of plant essential oils in dairy foods for the enhancement of shelf life (doctoral), probiotic-Indian dairy products (postdoctoral) and in the area of development of functional foods using micro/nano-colloidal delivery systems of potential plant bioactive components (postdoctoral). She is a well-recognized researcher has Web of Science Researcher ID: L-9405-2019; ORCID: https://orcid.org/0000-0001-6493-5217. She is in editorial board of few international journals including, *Medicinal Plants* – International Journal of Phytomedicines and Related Industries (ISSN: 0975-4261), *Microbial Biosystems* (ISSN 2357-0334), Egypt and *EUREKA: Life Sciences* (ISSN 2504-5695), Estonia. She is also the member of Society for Bioinformatics and Biological Sciences (SBBS). She is Visiting Assistant Professor, Department of Botany at Sri Avadh Raj Singh Smarak Degree College, Bishunpur Bairiya, Gonda (affiliated to Dr. RML Avadah University, Ayodhya, UP, India, (July 2019–); and Visiting Faculty, Department of Food Technology at Akal College of Agriculture, Eternal University, Baru Sahib, Himachal Pradesh, India (February 2020). She has published good few peer reviewed papers in internationally well reputed journals.

Jieyin Chen worked in the Institute of Food Science and Technology, Chinese Academy of Agricultural Sciences (2013 January–2020 March) and is currently working as professor at The Institute of Plant Protection, CAAS. He devotes himself on the research of phytopathogen, food quality and safety, and, has published over 50 papers.

Chapter 1
Fungal Byproducts in Food Technology

Shubhi Singh and Smriti Gaur

1.1 Introduction

Fungi are one of the most unique kingdoms in biology. Because of their distinctive characteristics, this kingdom has been a topic of intense studies by the researchers. Since fungi are known to have special feature to grow under extreme conditions and grow rapidly, these organisms have long been used at industrial level for the benefits of humans in their everyday life (Guerriero et al. 2015). Fungi have many beneficial roles in the field of economy, human health, biodecomposting, pharmaceuticals, and many more. One important facet regarding the use of fungal cells for different applications is the study of fungal cells' physiology. This term refers to study of the patterns of cell's nutrition, growth, reproduction, and metabolic activities in detail. Since fungal physiology plays a vital role in determining how fungus will interact with its biotic and abiotic environment, it will influence its response during its use in many applications (Walker and White 2017). Nowadays, fungi have been largely used in the area of food science, pharmaceuticals, and biodecomposting. During biodecomposting fungi are preferred more than the bacteria because these organisms have some special enzymes specifically for degrading the recalcitrant components. The recalcitrant components may be some phenolic compounds, azo dyes having the nitrogenous bonds, hydrocarbons, etc. (Tortella et al. 2015). Fungi are known to have rigid cell wall structure. As a result, this rigid cell wall protects the inner cell from some inhibitory components such as getting absorbed, giving this organism more versatility in resisting the harmful components (Gow et al. 2017). Since these organisms are eukaryotes in nature, they are known to contain a higher number of genes which provide them better ability to survive under any extreme conditions. The role of fungi in the field of pharmaceuticals is very significant. At

S. Singh · S. Gaur (✉)
Department of Biotechnology, Jaypee Institute of Information Technology, Noida, India
e-mail: smriti.gaur@jiit.ac.in

© The Author(s), under exclusive license to Springer Nature
Switzerland AG 2021
X. Dai et al. (eds.), *Fungi in Sustainable Food Production*, Fungal Biology,
https://doi.org/10.1007/978-3-030-64406-2_1

1

the industrial level, some special types of fungi species are usually grown for the production of many important enzymes, antibiotics, and vaccines (Mehdi et al. 2018). Moreover, considering the relationship of fungi with the field of food biotechnology, fungi have many pivotal contributions since ages. In the past decades, the use of fungi has been greatly prompted to make different kinds of food products. Fungi in food technology are used to make certain cultured food products which are formed from fermentation process performed by fungi, for example, yeast, or a part of fungus is consumed as a whole, for example, mushroom. Fungi are very particular in performing their metabolic processes. Due to these processes, the fungi are able to synthesize a wide variety of food products, medicines, vaccines, etc. These fungal products share great position in the biotechnology market. Apart from these fungal products which are produced by general procedures, there are some fungal byproducts too which are synthesized side by side. Advancements in technology have led to more research on these fungal byproducts. As a result, these byproducts when produced were extracted and further processed. They on further explorations were either converted into functional foods or are used in other food products as an active ingredient. With increasing awareness and health-related concerns, nowadays people are more inclined toward ingenious health-stimulating products. This chapter talks about the role of fungi in food biotechnology. Moreover, it includes information on several fungal byproducts, how they are produced, what their benefits are, and how these byproducts are further consumed in the field of food biotechnology. As the explorations on fungal products and byproducts are only partially conducted, the chapter also includes the future prospects of fungi in food biotechnology.

1.2   Use of Fungi in Food Biotechnology

Since millennial years, humans have struggled a lot for survival. As a consequence of this, they have discovered a variety of food products for their consumption. This discovery also includes the utilization of enormous range of organisms into the food products just to enhance the sensory or health beneficial properties of the products. The use of fungi in the food biotechnology is noteworthy. They are consumed either directly or are used as an additive just to carry out further food processing techniques for the production of specific food products. Fungi are considered to be very nutritive since they are good producers of some proteins and high-value biochemicals (Ojediran et al. 2016). These eukaryotic organisms are consumed in the food biotechnology in the form of fermentative yeasts, mushrooms, and filamentous fungi. The very first case of using the wild fungi for human consumptions was reported in China. Since then its use has been continued in different forms (Huang et al. 2015). As fungi have been known to have rich nutrition profile, its cultivation has been expanded in the past years just to meet the demands by the consumers. Mushrooms (macrofungi) are generally raised at large-scale industrial levels. Although many mushrooms species exist in nature, not all are edible. There are only few mushroom species which are cultivated for human consumption. The most popular mushroom which is used widely by humans is *Agaricus bisporus* or the button mushroom (Das and Arora 2018). Apart from this, there are other mushrooms too which share a good marketplace. For instance, shiitake mushrooms, truffles, morels, milk mushrooms, etc. have been used by humans in their variety of dishes. They are known to have high protein content and rich in antioxidant levels, vitamins, fiber, carbohydrates, and minerals (El Sebaaly et al. 2019). In addition to these, mushrooms are also known to contain lower cholesterol, fat, and caloric content. These health-promoting properties of mushrooms and their consumption have led to a development of a well-balanced diet. Many studies reveal that mushroom could act as an adjuvant in promoting health benefits. Since the nutrition profile of mushroom is very rich, they can be used as a source of prebiotics and can initiate the growth and regulation of human gut microbiota. These gut microbiota are known as probiotics which in return cause many health benefits (Ma et al. 2018). Apart from these, some reported data also reveals that mushrooms are a good source of many bioactive compounds (Beelman et al. 2019). These bioactive compounds are terpenes, phenolic compounds, and polysaccharides which are mainly responsible for enhancing the immune system functioning of an individual. Because of the presence of such components, mushrooms are good at exhibiting antioxidative, anti-inflammatory, and antimicrobial properties. As a result, mushroom extracts have also been used not only in food but also in cosmetics preparations (Taofiq et al. 2016).

The other form in which fungi are consumed is yeast. Yeasts are usually consumed because of their capacity to perform the process of fermentation, their potential to impart peculiar flavor to the food product, and their easy availability at a cheaper price (Ludlow et al. 2016). But for carrying out the fermentation procedure, only a particular type of yeast strain is selected. The criteria depend upon its ability to remain viable under processing conditions, ability to produce gas during fermentation, and ability to impart a desired flavor to the food product (Shurson 2018). The usage of yeasts can be seen during brewing, bread making, and preparation of many food items at domestic levels. The most used strain for the above-mentioned purpose is *Saccharomyces cerevisiae*. During preparation of these items, yeasts are known to produce carbon dioxide and alcohol as their byproducts. In the course of brewing, carbon dioxide (CO_2) is removed during the anaerobic fermentation process so that no other undesirable product gets formed except the alcohol (Walker and Stewart 2016). While in bread making, yeast ferments the available sugar to produce CO_2 and alcohol. The CO_2 produced is responsible for elasticity of the gluten protein present in the flour; as a result, the dough rose and the alcohol which was formed is evaporated during baking (Ali et al. 2012). In addition to these products, yeast cells have also been used in the dairy industry for the preparation of many fermented milk drinks, different varieties of cheese, etc. Yeasts are used not only for fermentation processes, but they are also being used to impart flavor and texture or initiate maturation of many products (De la Cerda and Bond 2019).

In addition to these two types of fungi, the filamentous fungi are also used for consumption by humans – not always in raw form but as an additive to many food products. For instance, many famous filamentous fungi like *Rhizopus oligosporus,*

Aspergillus oryzae, Monascus spp., etc. are widely utilized to produce the very famous fermented rice-based food product (Lv et al. 2015). Apart from fermenting the rice, these filamentous fungi are known to produce several other important secondary metabolites too which show potential health beneficial properties (Chen et al. 2015). The above-mentioned filamentous fungi are usually involved in the preparation of famous food products like soy sauce, miso, and tempeh. They are very old and prominent dishes in Asia. Soy sauce and miso are generally made using *Aspergillus oryzae*, where this fungus is known to continue fermentation for months to produce the desirable product (Lee et al. 2014). *Rhizopus oligosporus* is used in the preparation of tempeh. During its preparation, cooked and soaked soybeans are taken and then these are inoculated with fungus. This initiates the fermentation process and after 24–48 hours (may vary), the product gets ready for consumption (Tahir et al. 2018).

Besides these applications of fungi in the food industry, they are also known to excrete some useful enzymes while performing their metabolic activities. These enzymes are known to perform many important functions in food processing techniques. To illustrate, fungal lipases have been used for the flavor development in cheese, butter, and milk products and for removal of fat from fish and meat foodstuff (Singh and Mukhopadhyay 2012). Another important enzyme is fungal pectinase. This enzyme is produced from the famous fungi *Aspergillus niger* and holds a great place in the food industry. The main use of pectinase enzymes is in the extraction of fruit juices from its pulp followed by its clarification (Tapre and Jain 2014). Proteases extracted from fungal sources are also known to have significant contributions during brewing and in the preparation of baking food products (Srilakshmi et al. 2015). In addition to these, there are many other fungal enzymes too like lactases, pullulanases, hemicellulases, glucanases, α-amylases, β-amylases, etc. which play a crucial role in food processing techniques (Kavanagh 2017). Since the use of fungi directly or indirectly in the food industry has been discussed briefly, there are some fungal byproducts which are produced as well as consumed by humans and share a good nutraceutical and economic value.

1.3 Fungal Byproducts

During the processing techniques, fungus usually makes multiple byproducts. These products are then often harvested and further processed at the industrial levels and are consumed majorly as part of the human diet. Fungi are known to be the chief producers of various byproducts which are very significant to the food industry. These byproducts are fungal chitosan, fungal biomass, fungal proteins, fungal biochemicals, and many more. This section of the chapter talks about the various fungal byproducts, their biosynthetic methods, their importance, and how these products are used up as a part of the food industry.

At industrial levels, these fungal byproducts are synthesized on cheap and easy available substrates so that their production costs remain low. Further the harvested byproducts are processed prior to their use in the food industry. The figure below describes the various fungal byproducts which are discussed in this chapter in the next sections (Fig. 1.1).

1.3.1 Proteins from Filamentous Fungi Biomass

As discussed, fungi have a variety of significant applications in many fields. This eukaryotic organism is not only directly or indirectly consumed in the food sector but is also used widely to degrade different types of wastes. It could be solid, semi-solid, or liquid wastes (Eichlerová et al. 2015). There are many important species of fungi which are involved in various decomposition processes. During decomposition of wastes, fungi are known to produce fungal biomass. This fungal biomass is extracted and is further processed for the extraction of proteins (Asadollahzadeh et al. 2018). There are many factors which affect the production of biomass such as the type of bioreactor used, the type of substrates used, the organism used, operation parameters, and downstream processes (Reihani and Khosravi-Darani 2019). Earlier studies reveal that during crucial times of world wars, there was a need to feed people with a rich diet, to avoid mortality from malnutrition. Hence to the rescue, the fungal biomass was further processed to extract proteins. The proteins were then used as a food component in order to meet the nutritional needs (Suman et al. 2015). The source of substrates for growing fungal biomass matters a lot. Since the production happens at the industrial level, its cost is a crucial factor. For conducting the production at cheaper costs, the food processing wastewaters are the best way to grow the fungal biomass. Besides decomposition process, fungal biomass can also be produced from the fermentation process involving the use of fungi. Depending

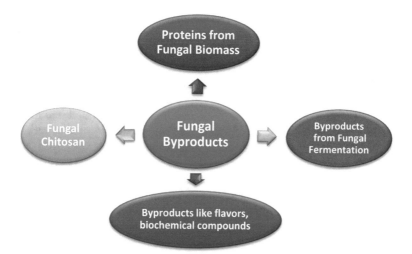

Fig. 1.1 Various types of fungal byproducts used in the food sector

upon the nature of substrate involved, the fermentation process producing the fungal biomass could be submerged fermentation or solid state fermentation. As mentioned above, the type of substrate is a pivotal factor from economic as well as from yield point of view. As many agro-industrial based-wastewaters are used as substrates, few more components must be added to the media to meet the nutritional demands of the fungi. The important components include carbon source, nitrogen source, and some other salts. Nitrogen in the media will regulate the pH conditions. Hence, its concentration should be accurate. Carbon source must be rich in cellulosic material during the fermentation process. But since cellulosic material is used, some kind of pretreatment must be required in order to make these nutrients available to the fungi (Hultberg and Bodin 2017).

Many studies have been conducted where researchers have used a variety of different substrates for the production of fungal biomass. One such study demonstrated the use of fungi *Trichoderma harzianum* for the production of fungal biomass. The substrate used was polished rice and it was found that the biomass produced maximum yield of 5% (w/v) and was rich in protein content with the presence of essential amino acids (Ahmed et al. 2017). Another study involved the combination of three white rot fungi for the production of fungal biomass through the process of fermentation. The substrates were also used in combination. The substrates chosen were from the agro-industrial wastes, i.e., fruit peels. In this study, they used pineapple, banana, and papaya peels. The fungi were inoculated and after fermentation and harvesting, the amount of fungal biomass obtained was measured (Saheed et al. 2016). Moreover, some used another type of fungi for high yield production of fungal biomass. Another study was done where researchers used various strains of *Morchella* species with the combination of substrates that were rich in starch. The substrates consumed were wheat grains and potato peels. The resulting biomass had high polysaccharide content with high yield (Papadaki et al. 2019).

Once the fungal biomass is produced, these fungal proteins were extracted and converted to more processed forms. The most common form in which these fungal biomass proteins are consumed nowadays is mycoprotein. Mycoproteins are also known by their other name – Quorn. The most common fungi used for mycoprotein production is *Fusarium venenatum* (Park et al. 2017). The usual and very first protocol for the production of this mycoprotein involves the inoculation of fungus into the glucose-containing media and with other controlled parameters. The incubation is carried out until a desired amount of circulating solids is formed into the air-lift fermenters. Once the solid mass is formed, the material is subjected to various sessions of heating so that RNA gets inactivated which is further removed from the cells through diffusion. The remaining solid mass is subjected to centrifugation so that clarification can be done. The last step is to perform the vacuum chilling and finally mycoprotein gets ready. But this protein further needs to get processed for its use as an edible component. Since the mycoprotein produced lacks elasticity, it was mixed with some common binding agents like egg albumin, water, or some other extracts. These binding agents are then mixed completely, and this modified mycoprotein mass is subjected to heating so that the egg proteins denature. After this, the modified mycoproteins are pressed in order to give them block or cube shapes and a textured appearance. As a result, the final modified form is ready to be consumed and is marketed under a brand name Quorn (Garodia et al. 2017). Nutritionally Quorn is very rich. This product is used as a substitute of meat by the vegetarian people. The texture, appearance, and flavor of Quorn are very much similar to meat but it is known to contain high nutrition profile. As the name suggests, this significant food product has a high amount of proteins with concentration of 11.5 g/100 g. It also contains high amounts of fibers and lower amounts of fats and cholesterol. The high amount of fibers results in the initiation of a feeling of fullness by releasing some satiety hormones and hence promotes good health (Finnigan 2011). All these qualities of this food component make it a crucial element in order to support food and protein security situations around the world.

1.3.2 Byproducts from Fungal Fermentation

While performing the fermentative processes to make the desired products, fungi are known to produce several byproducts. Since the production costs are very high, the market value of these products is low in the food sector. The common byproducts which are produced during the fermentation processes performed by fungi are various organic acids, glycerol, different kinds of enzymes, and fungal biomass. These different organic acids are fumaric acid, malic acid, citric acid, etc. They are produced in between the fermentation process and as the end product of fermentation. Generally there are two methods for the production of these organic acids, i.e., either by following the chemical processes or by using the fermentation technology. These organic acids are of great importance in the food industry. Every single acid is significant in its own way. During the mass production of organic acids, there are various kinds of fungi involved. To illustrate, many species of filamentous fungi *Aspergillus* have been used for the production of organic acids (Yang et al. 2017). In addition to this, *Rhizopus oryzae* has also been found to give good amounts of organic acids as the byproducts of their fermentation processes (Zhang et al. 2015). Yeasts like *Candida* and *Hansenula anomala* have been reported to be beneficial too in the production of these acids (Naraian and Kumari 2017). During their production at industrial levels, the types of substrates used must be accurate. Usually the substrates are consumed from the wastes produced by agro-industries, which are very much rich in carbon content. The media used for fungal growth have optimized culture as well as physical parameters. The production of organic acids can be divided into three main parts: preparation, inoculation, and fermentation and at last the final product can be harvested. Every individual organic acid is produced by following these three steps and the only difference lies in the type of raw materials, substrates, and fungi used. For instance, during the production of important and most widely used component in the food industry – citric acid – the starchy and cellulosic substrates are taken and solid state fermentation mode is preferred (John et al. 2012). The chemical and physical parameters must be optimized in order to achieve the highest yield. Once the acid is made, it is harvested and further processed in order to use it in foodstuffs. Citric acid in the food industry has many applications. The main use of this element is to adjust the pH (Yin et al. 2019). In case of alcoholic beverages like wines, beer, and whiskeys, citric acid is added not only to adjust pH but also to prevent turbidity (Chen et al. 2019). In carbonated beverages and juices, it is added to provide tartness and initiate the natural flavor of the juice (Ciriminna et al. 2017). Moreover, this component is also used in ice creams and milk-based products as an emulsifier (Euston and Goff 2019). It is also used to prevent the crystal formation of some sugars during the manufacturing of candies. Moreover, it is used as a preservative as it deactivates some oxidizing enzymes and prevents rotting of frozen fruits and vegetables (Mirza and Kasim 2019). The other important organic acid produced as a byproduct of fungal fermentation is gluconic acid. This has few applications in the food industry. This acid is used to impart particular sour taste to food products. It has the ability to improve the structure of dough during bread making (Yegin et al. 2018). Gluconic acid is believed to absorb lower fat during deep frying of certain foodstuffs (Alvarez-Sala et al. 2018). At the time of production, famous fungi *Aspergillus niger* are used to carry out fermentation. Its production at industrial levels includes fed-batch fermenters with optimized pH and temperature range. In addition to this, the substrates which are to be consumed must be highly rich in glucose (Ramachandran et al. 2017). Fumaric and lactic acids play a pivotal role in the food sector too. Many applications of the fumaric acid are the same as that of citric and gluconic acids. But it has also been found that fumaric acid is consumed in the form of artificial vinegar just to add flavors to the food products in the form of seasonings. During baking, it is sometimes used as a coagulant during the preparation of batter (Shukla et al. 2017). The contributions of lactic acid to the field of food technology are commendable. It is not only used as a flavoring agent, but it is also used to initiate formation of many milk-based products and pickles. Lactic acid has also been consumed as an acidulent and a preservative in the food industry. Though *Rhizopus* spp. have been used in the production of lactic acids at industrial levels, still there are many researches being conducted in order to enhance lactic acid production (Chen et al. 2018).

In addition to these organic acids, glycerol is produced as a byproduct of fungal fermentation. There are many factors affecting the production of glycerol during the alcoholic fermentation process. The amount of carbohydrates present, temperature, and the duration of fermentation decide how much glycerol will be produced (Mahboubi et al. 2017). More researches on the utilization of glycerol into the food sector have led to its production at the industrial level. Many strategies have been exploited to produce this component with maximum yield. Some used high osmotic salts with different concentrations in order to maximize its yield. It was seen that glycerol production gradually increased upon increasing the concentration of salts like NaCl and KCl (Petelenz-Kurdziel et al. 2013). Other studies demonstrated the use of two fungal species utilized as co-inoculation with yeast *Saccharomyces cerevisiae* which in return upregulated the oxidative balance pathway and resulted in the accumulation of more glycerols (Jiang et al. 2018). Glycerol has many significant applications in the food sector. It can either be used as a preservative or as a sweetener in foodstuffs (Johnson et al. 2017). Many studies show the use of glycerol as a thickening agent in any beverages. Since the nutritional profile of glycerol suggests that it contains low glycemic index, its use as an alternative to table sugars is preferred more (Balder 2018).

1.3.3 Fungal Chitosan

Chitosan is made up of many acetylated and deacetylated saccharide units. This component is present in the outer cell walls of the fungi and is known to impart rigidity to fungal cells. The other sources where chitosan can be extracted are from crustaceans, some filamentous fungi, yeasts, and algal sources. However, extraction of chitosan from crustaceans has many side effects such as the incorporation of allergic proteins along with chitosan, the availability of the sources, the regulatory affairs, and many more. Hence, the production of this component has been made possible from the fungal and algal sources. Fungal sources are usually considered for the extraction. As it is the major element of fungal cell walls, for its production, the fungal cells are cultivated, and later on, chitosan is harvested using some particular enzymes. It has been seen that for cultivation of fungal colonies, many cheap substrates are being used, especially the biowastes from the agro-based industries. This alternative is cheap and results in unlimited production of fungal species. One particular study demonstrates the use of popular enzyme – α-amylases – in order to harvest the chitosan from fungal cell walls. The use of this enzyme resulted in the higher yield of chitosan as compared to regular extraction processes (Ghormade et al. 2017). The study conducted by Tajdini and others in 2010 showed that chitosan could be extracted from the fungal mycelium too. During the study, spore suspensions from two fungal species *Rhizomucor miehei* and *Mucor racemosus* were used, and the applied culture conditions were continued till the particular amount of fungal mycelia is formed. It is followed by the separation of chitosan from mycelia by filtration and washing procedures. Then the pH was adjusted to neutral and purification along with deacetylation of extracted chitosan was performed. Another study also demonstrated the use of different fermentation procedures for chitosan production. In this study, the researchers have checked the maximum chitosan production by using both submerged and solid state fermentation with different nitrogen sources. It was concluded that urea was examined as the best nitrogen source during the fermentation process (Nwe et al. 2011). It was concluded from this study that submerged fermentation was able to produce more fungal biomass as well as fungal chitosan. The molecular weight of the harvested chitosan was also determined using the gel permeation chromatographic techniques, and it was further concluded that solid state fermentation is considered only for the extraction of lower molecular weight chitosan elements (Zhao et al. 2011). Moreover, another study states that cheap and effective production of chitosan could be made possible by just maintaining low pH and temperature conditions under unsterile conditions. All the culture media and physical parameters were optimized and validated during this study. It was concluded that under the optimized conditions, the production of chitosan was increased manifold (Tasar et al. 2016). Another recent study was conducted to produce the chitosan films. In this study, the authors incorporated fungal extracts from the famous edible fungi *Tricholoma*. It was seen that the addition of fungal extracts increased the elasticity and antioxidative and antimicrobial capacities of the chitosan films. However, there was a decrease in transparency and thermal stability of the final product (Koc et al. 2020).

Chitosan has many applications in the food industry. For the past many years, this component has been used as a clarifying agent in many beverages and fruit juices. To illustrate this, one such study describes the use of chitosan in clarifying the apple juice. The modified apple juice was more clarified without affecting the nutritional profile. Moreover, chitosan was proved to exhibit antioxidative and antimicrobial activities and hence it could be used as an active ingredient in the food industry (Abdelmalek et al. 2017). Apart from the above-mentioned properties, sometimes lower weight chitosan is used to encapsulate the bioactive components present in the food products. As a result, these components remain safe from enzymatic degradation during processing as well as from oxidative damage to certain extent. Furthermore, chitosan has the ability to form films and hence has a versatile application during food packaging and helps in the preservation of food products for longer periods of time (Wang et al. 2018). It also restricts the use of certain chemical preservatives to food products (Rocha et al. 2017). Chitosan can also be used to preserve minced meats for longer durations. One such study was done where the minced meat was exposed to fungal chitosan and to chemical preservative – potassium sorbate. It was noted that meat treated with chitosan showed higher antimicrobial activity as compared to meat treated with potassium sorbate. In addition to this, minced meat with chitosan exhibited greater sensory properties when compared to other one (Tayel et al. 2014).

1.3.4 Other Fungal Byproducts

Apart from the above-mentioned byproducts, fungi also produce various flavoring agents as their byproducts of fermentation process. Several compounds which are produced during the fungal fermentation processes are esters, volatile and nonvolatile acids, carbonyls, sulfur derivatives, and phenolic compounds. For instance, terpenes are the most abundant elements which are produced during the fungal fermentation. These elements are known to get produced from a variety of fungal species like *Trichoderma*, *Sporobolomyces*, *Kluyveromyces lactis*, etc. (Hermosa et al. 2014). Due to their production during fermentation period, the foodstuffs get particular flavor. Different terpenes are known to impart sometimes citric taste or sometimes fruity or bitter taste. Another component is lactone. Lactones are known to give fruity, nutty, buttery, or sometimes cheesy flavor to the foodstuffs. They are majorly seen in the foodstuffs related to dairy industry. Some particular dairy drinks, cheese, or edible dairy items have a specific and common taste and smell. This is because of a common compound – lactone – present in it. Usually *Trichoderma* and *Candida* fungi are used for the production of lactones (Siddiquee 2014). Pyrazines are also produced as byproducts of fungal fermentation. They impart roasty flavor to the food. They have an important role in the food industry as they are added additionally to the microwaved or canned food items to impart the additional flavor of roasting (Hu et al. 2019).

Besides imparting flavor to the foodstuffs, fungi are also known to produce various dyes as their byproducts. The colored pigments which are naturally produced as byproducts are melanins, flavins, carotenoids, quinones, etc. Carotenoids are largely produced by the zygomycetes (Dulf et al. 2020). Other pigments such as flavins, melanins, quinones, and anthraquinones are specifically extracted from the fungi *Fusarium* and *Drechslera* (Kumar et al. 2019). Since synthetic dyes/pigments used in many foodstuffs have many reported side effects, choosing the natural source for the extraction of these pigments was the safest alternative. These pigments are known to tolerate a wide range of pH and temperature. As a result of this, their applicability in food items plays a major role (Rao et al. 2017). The pigments mentioned above are used as food colorants on a large scale in the food industry. They are of great importance because of their unlimited and cheap production and easy extraction and they also have no harmful side effects and no seasonal variations (Aberoumand 2011). Riboflavin (vitamin B12) is also produced from fungi *Eremothecium ashbyii* (Subramaniam et al. 2020) which could be added to the food materials for enhancement of vitamin profile and also as a sunscreen to protect foodstuffs from damage caused by UV rays of the sun. Another recent study states that red-colored natural pigment lycopene could be extracted from fungus *Blakeslea trispora*. It has been also stated that *Monascus* could be used to extract several other food colorants and flavoring agents which can be used during the manufacturing of numerous flavored dairy products (Dufossé 2018).

Moreover, many enzymes are also produced from fungal sources as byproducts. To illustrate, fungal α-amylases are produced and extracted from many fungal species, majorly *Aspergillus oryzae*. *Trichoderma* are also able to synthesize many extracellular amylases. As a result, the cheap substrates like starch potato in soluble form were used to grow fungi, and then using extraction procedures, the enzymes were separated (Abdulaal 2018). Another study illustrates the production of fermentative liquor named as Chinese nong flavor liquor and the extraction of fungal α-amylases from the beverage. The presence of this enzyme was validated by checking the expression of amylase gene during starch-degrading process. It was concluded that amylase gene showed maximum expression while it was treated with starch-based compounds, amylopectin, glycogen, etc. (Yi et al. 2018). The fungal amylases have many applications in the food industry, for example, it can be used during the process of bread making. Amylases are known to hydrolyze starch into sugars, making it easier for the yeast cells to obtain their nutrition and grow. This also improves the final bread quality as its addition imparts some additional flavors to the dough too (Matsushita et al. 2020). Cellulases and proteases have been also found as byproducts of fungal enzymes. *Mucor* species like *Mucor pusillus* and *Mucor miehei* are the chief producers of proteases (Kour et al. 2019). The secreted enzyme shows the high proteolytic activity and hence can sometimes be used as a useful alternative to rennin during the production of cheese (Putranto et al. 2019).

1.4   Unenviable Fungal Byproducts

There are many other fungal byproducts which, when produced, impart some off-flavors to the food products. They are also known to affect the health negatively when consumed by the individuals. The most famous fungal byproducts which adversely ruin human health are mycotoxins. They are the secondary metabolites which are produced by the fungus during their metabolic processes. Apart from providing many health benefits and a variety of other food products, fungi are also famous for food spoilage (Avery et al. 2019). These spoilages of food items are caused by fungi when they produce certain mycotoxins or other unfavorable compounds. These mycotoxins have lower molecular weight and are known to cause several diseases. The general term used for diseases caused by mycotoxins is mycotoxicosis (Barac 2019). There are several factors which affect the growth of fungi and also the production of mycotoxins. These factors are temperature, water content present, amount of nutrients present, available amount of carbon and oxygen, etc. Stored food items often have these above-mentioned parameters in stabilized or fixed forms. In storage food items, when any contamination occurs, the microorganisms usually consume these available nutrients and release mycotoxins. The common mycotoxins which are produced from fungi are aflatoxins, ochratoxins, fumonisins, etc. (Bansal et al. 2011). Aflatoxins are very common among these and enters to the food chain through agricultural raw materials or from livestock. When water content exceeds more than its desired limit, the contaminant results in the growth of fungal molds. Usually dairy products, cheese, and breads after getting contaminated often have accumulated more aflatoxins. The source of contamination may be direct or indirect sometimes. Indirect contamination occurs when animals consume aflatoxicated feed and later the milk and related products when consumed from the same source cause many health-related problems (Dutton et al. 2012). Animals usually consume these toxins in B form which later on converts to M aflatoxins. These M aflatoxins are transferred to the milk. Literature reports the presence of M aflatoxins in the breast milk of mother, which can easily get transferred to the child (Ghiasian and Maghsood 2012). Many fungal species like *Aspergillus* and *Penicillium* are involved in the production of aflatoxins (Hammami et al. 2014). The use of contaminated raw materials from agricultural fields during the preparation of some foodstuffs results in the accumulation of these toxins. As these mycotoxins do not have any effect from heat or processing techniques, they remain unaltered. Moreover, once these mycotoxins enter the food chain, they do not get degraded by digestion or enzymatic attack. Mycotoxins are known to cause both acute and chronic diseases in humans. The degree of the illness depends upon many factors like intensity of exposure to the toxins, duration of exposure, age of an individual, and interaction of toxins with the immune system. These are known to affect the immune system by damaging macrophage functioning, inhibit the protein synthesis, and stop the ability of the lungs to clear minute foreign particles. Besides these mycotoxins, there are many off-flavors too which are synthesized from contaminated fungi. These fungi are responsible in the production of many volatile compounds which are mainly in charge of flavor and aroma development. Here also many fungi like *Penicillium, Botrytis cinerea, Trichoderma spp.,* etc. are responsible for the production of this volatile compound. The compounds are geosmin, MIB (2-methylisoborneol), alcohols like 2-heptanol, ketones like 1-octen-3-one, IPMP (2-isopropyl-3-methoxypyrazine, pyrazines), etc. These undesirable compounds are known to reduce the consumers' demand for the particular food items. The studies on preventing the development of these compounds into the food items are also limited. Hence, more research studies on the prevention and reducing their effects need to be conducted (Rousseaux et al. 2014).

1.5 Conclusion and Future Prospects

Fungi have many specific properties which make them an integral component in healthcare applications. This chapter describes the direct use of fungi in the food industry, the importance of secreted byproducts from fungi, their production strategies at industrial levels, and their importance when incorporated into the foodstuffs. Fungi have got the most attention from the research point of view. These organisms have shown many positive aspects when involved in food making. The fungal byproducts have also got much attention from the researchers. Applications of fungi are broad as these are used as a whole edible component; during fermentation; as developers of flavors, textures, taste, and aromas; as dietary supplements; etc. Since consumers are getting inclined more toward the natural products rather than the ones which have chemical preservatives, the fungal byproducts have been greatly used as an alternative in many food items. The use of fungi into the field of food biotechnology must include some more interesting developments in the coming years. Although several fungal species have already been discovered till now, still some of them are only used in the food industries. The same species are again and again exploited to check the potential benefits exhibited by fungi. It is necessary to exploit more new fungal species with new beneficial properties. Also new strategies must be explored in order to increase the production of fungal biomass and harvesting of essential components at a cheaper price. In addition to this, the strategies should be developed to reduce the impact of contaminant fungi into the food stuffs.

Fungal byproducts being a new discovery in the field of food biotechnology surely helped humans. Fungal products have been seen as a possible solution to perpetuate health. New insights in the pioneering of fungal compound-based dietary supplements and their importance promise new paradigms of researches and developments.

References

Abdelmalek BE, Sila A, Haddar A, Bougatef A, Ayadi MA (2017) β-Chitin and chitosan from squid gladius: biological activities of chitosan and its application as clarifying agent for apple juice. Int J Biol Macromol 104:953–962

Abdulaal WH (2018) Purification and characterization of α-amylase from Trichoderma pseudokoningii. BMC Biochem 19(1):4

Aberoumand A (2011) A review article on edible pigments properties and sources as natural biocolorants in foodstuff and food industry. WJDFS 6(1):71–78

Ahmed S, Mustafa G, Arshad M, Rajoka MI (2017) Fungal biomass protein production from Trichoderma harzianum using rice polishing. BioMed Res Int:1

Ali A, Shehzad A, Khan MR, Shabbir MA, Amjid MR (2012) Yeast, its types and role in fermentation during bread making process-a. Pak J Food Sci 22(3):171–179

Alvarez-Sala A, Blanco-Morales V, Cilla A, Garcia-Llatas G, Sánchez-Siles LM, Barberá R, Lagarda MJ (2018) Safe intake of a plant sterol-enriched beverage with milk fat globule membrane: bioaccessibility of sterol oxides during storage. J Food Compos Anal 68:111–117

Asadollahzadeh M, Ghasemian A, Saraeian A, Resalati H, Taherzadeh MJ (2018) Production of fungal biomass protein by filamentous fungi cultivation on liquid waste streams from pulping process. Bioresources 13(3):5013–5031

Avery SV, Singleton I, Magan N, Goldman GH (2019) The fungal threat to global food security. Fungal Biol 123(8):555–557

Balder E, Healthier Choices Management Corp (2018) Processes and methods of manufacture of arecoline. US Patent 9,994,884

Bansal J, Pantazopoulos P, Tam J, Cavlovic P, Kwong K, Turcotte AM, Lau BY, Scott PM (2011) Surveys of rice sold in Canada for aflatoxins, ochratoxin a and fumonisins. Food Addit Contam 28(6):767–774

Barac A (2019) Mycotoxins and human disease. In: Clinically relevant mycoses. Springer, Cham, pp 213–225

Beelman RB, Kalaras MD, Richie JP Jr (2019) Micronutrients and bioactive compounds in mushrooms: a recipe for healthy aging? Nutr Today 54(1):16–22

Chen W, He Y, Zhou Y, Shao Y, Feng Y, Li M, Chen F (2015) Edible filamentous fungi from the species Monascus: early traditional fermentations, modern molecular biology, and future genomics. Compr Rev Food Sci Food Saf 14(5):555–567

Chen X, Wang X, Xue Y, Zhang TA, Li Y, Hu J, Tsang YF, Zhang H, Gao MT (2018) Influence of rice straw-derived dissolved organic matter on lactic acid fermentation by Rhizopus oryzae. J Biosci Bioeng 125(6):703–709

Chen H, Chen Y, Ye X, Liu D, Chen J (2019) Turbidity, antioxidant compounds, color, and dynamics of clarification of bayberry juice using various polysaccharide based clarifying agents. J Food Process Preserv 43(7):e13980

Ciriminna R, Meneguzzo F, Delisi R, Pagliaro M (2017) Citric acid: emerging applications of key biotechnology industrial product. Chem Cent J 11(1):22

Das I, Arora A (2018) Alternate microwave and convective hot air application for rapid mushroom drying. J Food Eng 223:208–219

de la Cerda R, Bond U (2019) Accelerated evolution of lager yeast strains for improved flavour profiles. Access Microbiol 1(1A):496

Dufossé L (2018) Red colourants from filamentous fungi: are they ready for the food industry? J Food Compos Anal 69:156–161

Dulf FV, Vodnar DC, Toşa MI, Dulf EH (2020) Simultaneous enrichment of grape pomace with γ-linolenic acid and carotenoids by solid-state fermentation with Zygomycetes fungi and antioxidant potential of the bioprocessed substrates. Food Chem 310:125927

Dutton MF, Mwanza M, de Kock S, Khilosia LD (2012) Mycotoxins in south African foods: a case study on aflatoxin M 1 in milk. Mycotoxin Res 28(1):17–23

Eichlerová I, Homolka L, Žifčáková L, Lisá L, Dobiášová P, Baldrian P (2015) Enzymatic systems involved in decomposition reflects the ecology and taxonomy of saprotrophic fungi. Fungal Ecol 13:10–22

El Sebaaly Z, Assadi F, Sassine YN, Shaban N (2019) Substrate types effect on nutritional composition of button mushroom (Agaricus Bisporus). Poljoprivreda i Sumarstvo 65(1):73–80

Euston SR, Goff HD (2019) Emulsifiers in dairy products and dairy substitutes. In: Food emulsifiers and their applications. Springer, Cham, pp 217–254

Finnigan TJA (2011) Mycoprotein: origins, production and properties. In: Handbook of food proteins. Woodhead Publishing, pp 335–352

Garodia S, Naidu P, Nallanchakravarthula S (2017) QUORN: an anticipated novel protein source. IJSRST 3:2395–6011

Ghiasian SA, Maghsood AH (2012) Infants' exposure to aflatoxin M1 from mother's breast milk in Iran. Iran J Public Health 41(3):119

Ghormade V, Pathan EK, Deshpande MV (2017) Can fungi compete with marine sources for chitosan production? Int J Biol Macromol 104:1415–1421

Gow NA, Latge JP, Munro CA (2017) The fungal cell wall: structure, biosynthesis, and function. In: The fungal kingdom, pp 267–292

Guerriero G, Hausman JF, Strauss J, Ertan H, Siddiqui KS (2015) Destructuring plant biomass: focus on fungal and extremophilic cell wall hydrolases. Plant Sci 234:180–193

Hammami W, Fiori S, Al Thani R, Kali NA, Balmas V, Migheli Q, Jaoua S (2014) Fungal and aflatoxin contamination of marketed spices. Food Control 37:177–181

Hermosa R, Cardoza RE, Rubio MB, Gutiérrez S, Monte E (2014) Secondary metabolism and antimicrobial metabolites of trichoderma. In: Biotechnology and biology of trichoderma. Elsevier, pp 125–137

Hu H, Liu H, Shi A, Liu L, Fauconnier ML, Wang Q (2019) The effect of microwave pretreatment on micronutrient contents, oxidative stability and flavor quality of peanut oil. Molecules 24(1):62

Huang Q, Jia Y, Wan Y, Li H, Jiang R (2015) Market survey and risk assessment for trace metals in edible fungi and the substrate role in accumulation of heavy metals. J Food Sci 80(7):H1612–H1618

Hultberg M, Bodin H (2017) Fungi-based treatment of brewery wastewater – biomass production and nutrient reduction. Appl Microbiol Biotechnol 101(11):4791–4798

Jiang C, Chen X, Lei S, Shao D, Zhu J, Liu Y, Shi J (2018) Fungal spores promote the glycerol production of Saccharomyces cerevisiae by upregulating the oxidative balance pathway. J Agric Food Chem 66(12):3188–3198

John KS, Ali MN, Umakumar G, Tabassum H (2012) Studies on inductive effect of methanol on production of citric acid from waste cellulosic substrates using locally isolated Aspergillus niger and MTTC Aspergillus niger strains. Int J Eng Sci Technol 4(2):431–441

Johnson W, Lee T, May RE, Concentrate Manufacturing Co of Ireland (2017) Non-nutritive sweetened beverages with glycerine. U.S. Patent application 15/213,790

Kavanagh K (ed) (2017) Fungi: biology and applications. Wiley

Koc B, Akyuz L, Cakmak YS, Sargin I, Salaberria AM, Labidi J, Ilk S, Cekic FO, Akata I, Kaya M (2020) Production and characterization of chitosan-fungal extract films. Food Biosci 35:100545

Kour D, Rana KL, Thakur S, Sharma S, Yadav N, Rastegari AA, Yadav AN, Saxena AK (2019) Disruption of protease genes in microbes for production of heterologous proteins. In: New and future developments in microbial biotechnology and bioengineering. Elsevier, pp 35–75

Kumar A, Prajapati S, Nandan S, Neogi TG (2019) Industrially important pigments from different groups of fungi. In: Recent advancement in white biotechnology through fungi. Springer, Cham, pp 285–301

Lee S, Seo MH, Oh DK, Lee CH (2014) Targeted metabolomics for Aspergillus oryzae-mediated biotransformation of soybean isoflavones, showing variations in primary metabolites. Biosci Biotechnol Biochem 78(1):167–174

Ludlow CL, Cromie GA, Garmendia-Torres C, Sirr A, Hays M, Field C, Jeffery EW, Fay JC, Dudley AM (2016) Independent origins of yeast associated with coffee and cacao fermentation. Curr Biol 26(7):965–971

Lv XC, Cai QQ, Ke XX, Chen F, Rao PF, Ni L (2015) Characterization of fungal community and dynamics during the traditional brewing of Wuyi Hong Qu glutinous rice wine by means of multiple culture-independent methods. Food Control 54:231–239

Ma G, Yang W, Zhao L, Pei F, Fang D, Hu Q (2018) A critical review on the health promoting effects of mushrooms nutraceuticals. Food Sci Human Wellness 7(2):125–133

Mahboubi A, Ferreira JA, Taherzadeh MJ, Lennartsson PR (2017) Production of fungal biomass for feed, fatty acids, and glycerol by Aspergillus oryzae from fat-rich dairy substrates. Fermentation 3(4):48

Matsushita K, Tamura A, Goshima D, Santiago DM, Myoda T, Takata K, Yamauchi H (2020) Effect of combining additional bakery enzymes and high pressure treatment on bread making qualities. J Food Sci Technol 57(1):134–142

Mehdi Y, Létourneau-Montminy MP, Gaucher ML, Chorfi Y, Suresh G, Rouissi T, Brar SK, Côté C, Ramirez AA, Godbout S (2018) Use of antibiotics in broiler production: global impacts and alternatives. Anim Nutr 4(2):170–178

Mirza SK, Kasim SS (2019) Determination of antioxidant and acidity regulators as additives in different fruit juices of India. Eur Chem Bull 8(7):235–238

Naraian R, Kumari S (2017) Microbial production of organic acids. Microbial Functional Foods and Nutraceuticals, p 93

Nwe N, Furuike T, Tamura H (2011) Production, properties and applications of fungal cell wall polysaccharides: chitosan and glucan. In: Chitosan for biomaterials II. Springer, Berlin/ Heidelberg, pp 187–207

Ojediran TK, Ogunmola BT, Ajayi AO, Adepoju MA, Odelade K, Emiola IA (2016) Nutritive value of processed dietary fungi treated Jatropha curcas (L). Kernel meals: voluntary intake, growth, organ weight and hepatic histology of broiler chicks. Trop Agric 93(2):101–110

Papadaki A, Diamantopoulou P, Papanikolaou S, Philippoussis A (2019) Evaluation of biomass and chitin production of Morchella mushrooms grown on starch-based substrates. Foods 8(7):239

Park HS, Jun SC, Han KH, Hong SB, Yu JH (2017) Diversity, application, and synthetic biology of industrially important Aspergillus fungi. In: Advances in applied microbiology, vol 100. Academic, pp 161–202

Petelenz-Kurdziel E, Kuehn C, Nordlander B, Klein D, Hong KK, Jacobson T, Dahl P, Schaber J, Nielsen J, Hohmann S, Klipp E (2013) Quantitative analysis of glycerol accumulation, glycolysis and growth under hyper osmotic stress. PLoS Comput Biol 9(6)

Putranto WS, Suhartono MT, Kusumaningrum HD, EGiriwono P, Mustopa AZ, Suradi K, Chairunnisa H (2019) Fresh cheese probiotic with local isolate Lactobacillus casei 2.12 as starter in fermentation. In: IOP conference series: earth and environmental science (vol 334, no 1, p 012048). IOP Publishing

Ramachandran S, Nair S, Larroche C, Pandey A (2017) Gluconic acid. In: Current developments in biotechnology and bioengineering. Elsevier, pp 577–599

Rao N, Prabhu M, Xiao M, Li WJ (2017) Fungal and bacterial pigments: secondary metabolites with wide applications. Front Microbiol 8:1113

Reihani SFS, Khosravi-Darani K (2019) Influencing factors on single-cell protein production by submerged fermentation: a review. Electron J Biotechnol 37:34–40

Rocha MAM, Coimbra MA, Nunes C (2017) Applications of chitosan and their derivatives in beverages: a critical review. Curr Opin Food Sci 15:61–69

Rousseaux S, Diguta CF, Radoï-Matei F, Alexandre H, Guilloux-Bénatier M (2014) Non-Botrytis grape-rotting fungi responsible for earthy and moldy off-flavors and mycotoxins. Food Microbiol 38:104–121

Saheed OK, Jamal P, Karim MIA, Alam MZ, Muyibi SA (2016) Utilization of fruit peels as carbon source for white rot fungi biomass production under submerged state bioconversion. J King Saud Univ Sci 28(2):143–151

Shukla P, Akshay S, Ashok S (2017) Food additives from an organic chemistry perspective. MOJ Biorgan Org Chem 1(3)

Shurson GC (2018) Yeast and yeast derivatives in feed additives and ingredients: sources, characteristics, animal responses, and quantification methods. Anim Feed Sci Technol 235:60–76

Siddiquee S (2014) Recent advancements on the role and analysis of volatile compounds (VOCs) from trichoderma. In: Biotechnology and biology of trichoderma. Elsevier, pp 139–175

Singh AK, Mukhopadhyay M (2012) Overview of fungal lipase: a review. Appl Biochem Biotechnol 166(2):486–520

Srilakshmi J, Madhavi J, Lavanya S, Ammani K (2015) Commercial potential of fungal protease: past, present and future prospects. J Pharm Chem Biol Sci 2(4):218–234

Subramaniam V, Das S, Sandeep K (2020) Production and partial characterization of chitinase and glucanase produced by riboflavin overproducer Eremothecium Ashbyii. Available at SSRN 3528076

Suman G, Nupur M, Anuradha S, Pradeep B (2015) Single cell protein production: a review. Int J Curr Microbiol App Sci 4(9):251–262

Tahir A, Anwar M, Mubeen H, Raza S (2018) Evaluation of physicochemical and nutritional contents in soybean fermented food tempeh by Rhizopus oligosporus. J Adv Biol Biotechnol:1–9

Tajdini F, Amini MA, Nafissi-Varcheh N, Faramarzi MA (2010) Production, physiochemical and antimicrobial properties of fungal chitosan from Rhizomucor miehei and Mucor racemosus. Int J Biol Macromol 47(2):180–183

Taofiq O, González-Paramás AM, Martins A, Barreiro MF, Ferreira IC (2016) Mushrooms extracts and compounds in cosmetics, cosmeceuticals and nutricosmetics – a review. Ind Crop Prod 90:38–48

Tapre AR, Jain RK (2014) Pectinases: enzymes for fruit processing industry. Int Food Res J 21(2)

Tasar OC, Erdal S, Taskin M (2016) Chitosan production by psychrotolerant Rhizopus oryzae in non-sterile open fermentation conditions. Int J Biol Macromol 89:428–433

Tayel AA, Ibrahim SI, Al-Saman MA, Moussa SH (2014) Production of fungal chitosan from date wastes and its application as a biopreservative for minced meat. Int J Biol Macromol 69:471–475

Tortella G, Durán N, Rubilar O, Parada M, Diez MC (2015) Are white-rot fungi a real biotechnological option for the improvement of environmental health? Crit Rev Biotechnol 35(2):165–172

Walker GM, Stewart GG (2016) Saccharomyces cerevisiae in the production of fermented beverages. Beverages 2(4):30

Walker GM, White NA (2017) Introduction to fungal physiology. In: Fungi: biology and applications, pp 1–35

Wang H, Qian J, Ding F (2018) Emerging chitosan-based films for food packaging applications. J Agric Food Chem 66(2):395–413

Yang L, Lubeck M, Lübeck PS (2017) Aspergillus as a versatile cell factory for organic acid production. Fungal Biol Rev 31(1):33–49

Yegin S, Altinel B, Tuluk K (2018) A novel extremophilic xylanase produced on wheat bran from Aureobasidium pullulans NRRL Y-2311-1: effects on dough rheology and bread quality. Food Hydrocoll 81:389–397

Yi Z, Fang Y, He K, Liu D, Luo H, Zhao D, He H, Jin Y, Zhao H (2018) Directly mining a fungal thermostable α-amylase from Chinese Nong-flavor liquor starter. Microb Cell Fact 17(1):30

Yin FW, Zhang YT, Jiang JY, Guo DS, Gao S, Gao Z (2019) Efficient docosahexaenoic acid production by Schizochytrium sp. via a two-phase pH control strategy using ammonia and citric acid as pH regulators. Process Biochem 77:1–7

Zhang K, Yu C, Yang ST (2015) Effects of soybean meal hydrolysate as the nitrogen source on seed culture morphology and fumaric acid production by Rhizopus oryzae. Process Biochem 50(2):173–179

Zhao LM, Shi LE, Zhang ZL, Chen JM, Shi DD, Yang J, Tang ZX (2011) Preparation and application of chitosan nanoparticles and nanofibers. Braz J Chem Eng 28(3):353–362

Chapter 2
Fungal Production of Dietary Fibers

Divya and Shashank Mishra

2.1 Definition

Dietary fiber or roughage is the indigestible portion of food derived from plants. It has two main components (Dietary Reference Intakes for Energy 2005):

- Soluble fiber, which dissolves in water.
- Insoluble fiber, which does not dissolve in water.

There are various definitions for dietary fiber such as:

1. Institute of Medicine – Dietary fiber consists of no digestible carbohydrates and lignin that are intrinsic and intact in plants. Functional fiber consists of isolated, nondigestible carbohydrates that have beneficial physiologic effects in humans. Total fiber is the sum of dietary fiber and functional fiber (Institute of Medicine 2005).

2. American Association of Cereal Chemists – Dietary fiber is the edible parts of plants or analogous carbohydrates that are resistant to digestion and absorption in the human small intestine, with complete or partial fermentation in the large intestine. Dietary fiber includes polysaccharides, oligosaccharides, lignin, and associated plant substances. Dietary fibers promote beneficial physiologic effects including laxation, and/or blood cholesterol attenuation, and/or blood glucose attenuation (American Association of Cereal Chemists 2001).

Divya
Department of Bio-Engineering, Birla Institute of Technology, Ranchi, Jharkhand, India

S. Mishra (✉)
Quality Control and Quality Assurance Laboratory, Biotech Park,
Lucknow, Uttar Pradesh, India

© The Author(s), under exclusive license to Springer Nature
Switzerland AG 2021
X. Dai et al. (eds.), *Fungi in Sustainable Food Production*, Fungal Biology,
https://doi.org/10.1007/978-3-030-64406-2_2

3. Codex Alimentarius Commission – Dietary fiber means carbohydrate polymers with ≥ 10 monomeric units, which are not hydrolyzed by the endogenous enzymes in the small intestine of humans (Codex Alimentarius Commission 2009).

2.2 Introduction

Fungi are ideal food because they have a fairly high content of protein (typically 20–30% dry matter as crude protein) which contains all of the essential amino acids essential to human and animal nutrition. Fungal biomass is also a source of dietary fiber, free of cholesterol.

Mushrooms are the main examples of fungal production of dietary fibers that are cultivated around the world, their global annual production being in the region of 8 million metric tonnes.

China is largest producer of mushrooms, nearly 60% of the world's mushrooms. Supplies of fresh mushrooms are now intercontinental goods. The only successful fermenter-grown fungal food on the market is the mycoprotein Quorn, the mycelium of a species of *Fusarium*. In marketing the material, its ability to simulate the fibrous nature of meat is emphasized and it is sold as a healthy alternative to meat. A large-scale collection of mushrooms for food has become an industry in many regions and the commercial picking industry is bound to continue to expand. However, it raises several issues, including conservation, ownership, and sustainability of supplies. The key reaction seems to be effective holistic management of the forest resource (Moore and Chiu 2001). Dietary fiber is a mixture of plant carbohydrate polymers, consisting of polysaccharides and oligosaccharides, as well as hemicelluloses, cellulose, resistant starch, pectin substances, insulin, and gums, among others (Fuentes-Zaragoza et al. 2010).

Our focus is on the fungal production of dietary fibers, but we must acknowledge that the yeasts play a dominant role where biotechnology is applied to the food industry, being essential in brewing and bread making, and important sources of single-cell protein and dietary supplements. In various industries, these fungal activities are essential to human existence and support. On the other hand, traditional solid state fermentations for producing mushrooms and other food products and in recent years the Quorn fermentation provide us with a sufficient range of examples of filamentous fungi being almost equally crucial to human affairs. Judging from archaeological and similar finds, mushrooms and bracket fungi have been used for both food and medicinal purposes (Moore and Chiu 2001).

Mushrooms are great example of fungal production of dietary fiber. They have valuable resources for food, medicine, and nutraceuticals. Edible mushroom is considered as a novel source of dietary fiber (Cheung 2013).

2.3 Production

Mushrooms or fungi are underutilized compared to other sources of dietary fiber such as cereals, fruits, legumes, and vegetables (O'Shea et al. 2012; Elleuch et al. 2010).

Fungi that have distinctive and visible fruiting bodies are defined as mushrooms (Chang and Miles 1992) and they include edible and medicinal ones. The flesh or dried form fruiting bodies of edible mushrooms (e.g., *Lentinus edodes*) are mainly consumed, while medicinal mushrooms (e.g., *Ganoderma lucidum*) are non-edible fungi that are used for biopharmaceutical purposes because they contain bioactive components such as polysaccharides and triterpenoids.

Mushroom cell walls enclose a mixture of fibrillar and matrix components which include chitin (a straight-chain $(1 \rightarrow 4)$-β-linked polymer of N-acetyl-glucosamine) and the polysaccharides such as $(1 \rightarrow 3)$-β-D-glucans and mannans, respectively (Bartnicki-Garcia 1970).

2.3.1 Preparation of Fungus (Mushroom) as Dietary Fiber

Dietary fiber preparation involves the removal of non-dietary fiber materials by either enzymatic or chemical methods. Soluble dietary fiber is extracted by water or other aqueous solution with pH control, while insoluble dietary fiber is improved as insoluble filtrate. The most widely accepted method for total dietary fiber determination is the AOAC enzymatic-gravimetric method 985.29 (AOAC International 2000) involving the use of three analytical enzymes: heat-stable α-amylase (EC 3.2.1.1), protease (EC 3.4.21.14), as well as amyloglucosidase (EC 3.2.1.3) to remove all non-fiber materials including starch and protein. Chemical methods for dietary fiber preparation are usually simple but non-specific.

Compared to mycelium, mushroom sclerotium has the highest level of nondigestible carbohydrates and an ideal source of commercial dietary fiber. Recently, an enzymatic procedure modified from the AOAC method 985.29 for preparing some novel dietary fibers from three mushroom sclerotia including *Pleurotus tuberregium* (PTR), *Polyporus rhinocerus* (PR), and *Wolfiporia cocos* (WC) using analytical and industrial food-grade enzymes was developed in laboratories (Wong and Cheung 2005, 2009). The effects of these enzymes on both the yield and nondigestible carbohydrate composition of sclerotial dietary fiber were compared. Use of industrial grade enzymes in the preparation the total dietary fiber gives a very high yield of sclerotial dietary fibers [PTR, 81.2%; PR, 86.5%; WC, 96.2% dry weight (DW)] with purity that was comparable to that of analytical enzymes (Wong and Cheung 2009; McCleary 2000). Additionally, in the application of food products, sclerotial dietary fiber has functional properties such as water- and oil holding, emulsifying, and mineral binding (Wong and Cheung 2005). In the food industry, dietary fiber has its own advantages such as energy saving, environmental friendly, non-toxic, and being specific (Cheung 2013).

2.3.2 Commercial Mushroom Production in the Asian Tradition

Lentinula edodes (mushroom) is another example of fungal production of dietary fiber, traditionally grown on evergreen hardwood logs (oak, chestnut, hornbeam) and is still very widely grown like this in the central highlands in China. The growing region covers an area equal to the entire land area of the European Union. The logs considered suitable for shiitake (*Lentinula edodes*) production are over 10 cm diameter and 1.5 to 2 m long. To minimize pre-infestation by wild fungi or insects, the logs are normally cut in spring or autumn of each year. Holes drilled in the logs (or saw or axe cuts) are packed with spawn, and the spawn-filled hole then sealed with wax or other sealant to protect the spawn from weather. To permit good air circulation and easy drainage and provide temperatures between 24 °C and 28 °C, the logs are stacked in laying yards on the open hillside. The logs are kept for 5 to 8 months to grow fungus completely through the log. Finally, the logs are transferred to the raising yard to promote fruit body formation. This procedure is done in winter to ensure the lower temperature (12–20 °C) and increased moisture which are required for fruit body initiation. The first crops of mushrooms appear in the first spring after being moved to the raising yard. Each log will produce 0.5 to 3 kg of mushrooms, each spring and autumn, for 5 to 7 years. On both land and trees, the traditional approach to shiitake production is expensive and demanding. Several other agricultural wastes make suitable substrates for mushroom production of dietary fibers such as the straw mushroom (*Volvariella volvacea*) grown mainly on rice straw. For the preparations of the substrate, straw bundles are soaked in water for 24 to 48 hours. The soaked straw is piled into heaps about 1 m high which are inoculated with spent straw from a previous crop. In less than 1 month, a synchronized flush of egg-like fruit bodies appears. These immature fruit bodies are sold for consumption just like the young fruits. Comparatively low yields of straw mushroom (*Volvariella volvacea*) are generated from the substrate, and it is difficult to maintain a good quality in post-harvest storage. Within 2–3 days the crop turns brown and autolyses even in cold storage. These factors restrict the production of the crop (Moore and Chiu 2001).

2.4 Importance

In human bodies dietary fiber plays a very important role for regulation (Anderson et al. 2009). Dietary fiber is not able to be decayed in the human gut and affects the moisture absorption in the digestive system. It increases the volume of food inside the intestines and stomach, increases satiety, and facilitates weight loss (Manzoni et al. 2008). Dietary fiber promotes gastrointestinal peristalsis to alleviate constipation (Tse et al. 2000) and absorbs harmful materials in the gut, for removal (Borycka 2010; Ta et al. 1999). For probiotics proliferation, dietary fibers improve the intestinal flora and provide energy (Wu et al. 2011). Studies shows that dietary fiber helps to reduce postprandial blood glucose, insulin, and triglyceride concentrations (Giacco et al. 2002; Maand and Mu 2016) and can lower blood cholesterol levels (Cheung 2013; Jimenez-Escrig and Sanchez-Muniz 2000).

Fungal biomass is easily digested, the chitinous wall provides a source of dietary fiber, and although filamentous fungi, in contrast to yeasts, have relatively low vitamin content, they do contain B vitamins and are characteristically low in fat. Also, an enormously significant quality of all fungal food is that it is practically free of cholesterol. Subsequently, fungal protein foods compete successfully with animal protein foods (i.e., meat) on health grounds. Since, in principle, fungal foods can be produced readily using waste products as substrates, fungal foods should also be able to compete successfully on grounds of primary cost (Moore and Chiu 2001).

2.5 Conclusion

Application of fungal production on dietary fiber (mushroom) as efficient food ingredient or product provides various health benefits (Cheung 2013). Based on existing studies, it is clear that dietary fiber is a significant component of diet and nutrition. Studies have analyzed a number of health benefits associated with dietary fiber, including the promotion of healthier bowel function, lowering the level of cholesterol in the body, and controlling blood sugar levels. Studies showed that maximum fungal production of dietary fiber is acceptable in our society as food or medicine. From observation from various literatures, it is shown that dietary fiber plays a critical role in the shelf-life of dietary products. Lots of health benefits and positive effects on shelf-life provided by dietary fiber are important and justify significant further research (Anderson et al. 2009).

2.6 Future Prospects

Chapter deals with fungus as food and how food industries develop in the future. Fungi have very significant roles in processing or as components of food in supplementary nutrients and health products. Our focus is on the filamentous fungi but we should acknowledge yeasts also because they play a dominant role in the food industry such as in brewing, in bread making, and as important sources of single-cell protein and dietary supplements. These fungal activities are crucial to human survival and need huge support. In recent years, traditional solid state fermentations for producing mushrooms and other food products and the Quorn fermentation provide us with a sufficient range of examples of filamentous fungi that are more crucial to human affairs. Archaeological and similar finds, mushrooms, toadstools, and bracket fungi are recorded in history for their both food and medicinal purposes. To grow and sell fungal dietary fiber as mushrooms in both developed and developing countries, cultivation industry is structured in such a way that allows all of its benefits to be returned to the mushroom farmers around the world. Biotechnology includes the terms "market pull" (determination of product success by market demand and the positive response of consumer) and "science push" (result of the scientific innovation, where there is an introduction of innovative products to the market). Exploitation of filamentous fungi must be considered for community impact of exploitation in future development. We should apply cultivation of mushroom with waste disposal and remediation integrated with the latest research and traditional methods for developing cultivation methods, so that the industry is sustainable with this integration. In fermentation industries, the real value of the product should far outweigh any adverse environmental impacts of the industrial process itself or its wastes. Above all, there is need for wider and deeper research into the biology of fungi. Literature surveys reveal ignorance more precisely than our knowledge (Moore and Chiu 2001).

References

American Association of Cereal Chemists (2001) The definition of dietary fiber: report of the Dietary Fiber Definition Committee to the Board of Directors of the American Association of Cereal Chemists. Cereal Foods World 46:112–126

Anderson JW, Baird P, Davis RH Jr et al (2009) Health benefits of dietary fiber. Nutr Rev 67(4):188–205

AOAC International (2000) Official methods of analysis of AOAC International, 17th edn, Arlington

Bartnicki-Garcia S (1970) Cell wall composition and other biochemical markers in fungal phylogeny. In: Harborne JB (ed) Phytochemical phylogeny. Academic, London, pp 81–103

Borycka BKR (2010) Binding cadmium and lead using natural polysaccharide fibers from some fruit and vegetable wastes. Zywnosc: Nauka, Technologia, Jakosc 69(2):104–110

Chang ST, Miles G (1992) Mushroom biology: a new discipline. Mycologist 6:64–65

Cheung PCK (2013) Mini-review on edible mushrooms as source of dietary fiber: preparation and health benefits. 2(3–4):162–166

Codex Alimentarius Commission (2009) Food and Agriculture Organization; World Health Organization. Report of the 30th session of the Codex Committee on nutrition and foods for special dietary uses. ALINORM 9/32/26. Cited 2012 Mar 27

Dietary Reference Intakes for Energy, Carbohydrate, fibre, Fat, Fatty Acids, Cholesterol, Protein, and Amino Acids (Macronutrients) (2005) Chapter 7: Dietary, functional and total fibre. US Department of Agriculture, National Agricultural Library and National Academy of Sciences, Institute of Medicine, Food and Nutrition Board

Elleuch M, Bedigian D, Roiseux O et al (2010) Dietary fiber and fiber-rich by-products of food processing: characterization, technological functionality and commercial application: a review. Food Chem 124:411–421

Fuentes-Zaragoza E, Riquelme-Navarrete MJ, Sanchez-Zapata E, PerezAlvarez JA (2010) Resistant starch as functional ingredient: a review. Food Res Int 43:931–942

Giacco R, Clemente G, Riccardi G (2002) Dietary fibre in treatment of diabetes: myth or reality? Digest Liver Dis 34(Supplement2):S140–S144

Institute of Medicine (2005) Food and Nutrition Board. Dietary reference intakes: energy, carbohydrates, fiber, fat, fatty acids, cholesterol, protein and amino acids. National Academies Press, Washington, DC

Jimenez-Escrig A, Sanchez-Muniz FJ (2000) Dietary fibre from edible seaweeds: chemical structure, physicochemical properties and effects on cholesterol metabolism. Nutr Res 20(4):585–598

Maand M, Mu T (2016) Anti-diabetic effects of soluble and insoluble dietary fibre from deoiled cumin in low-dose streptozotocin andhighglucose-fatdiet-inducedtype 2 diabeticrats. J Funct Food 25:186–196

Manzoni GM, Castelnuovo G, Molinari E (2008) Weight loss with a low-carbohydrate, Mediterranean, or low-fat diet. New Engl J Med 359(20):2170–2172

McCleary BV (2000) Importance of enzyme purity and activity in the measurement of total dietary fiber and dietary fiber components. J AOAC Int 83:997–1005

Moore D, Chiu SW (2001) Fungal products as food. In: Pointing SB, Hyde KD (eds) Chapter 10 in Bio-exploitation of filamentous fungi. Fungal Diversity Press, Hong Kong, pp 223–251

O'Shea N, Arendt EK, Gallagher E (2012) Dietary fibre and phytochemical characteristics of fruit and vegetable by-products and their recent applications as novel ingredients in food products. Innovative Food Sci Emerg Technol 16:1–10

Ta CA, Zee JA, Desrosiers T et al (1999) Bindingcapacityofvariousfibretopesticideresiduesun dersimulatedgastrointestinal conditions. Food Chem Toxicol 37(12):1147–1151

Tse PWT, Leung SSF, Chan T, Sien A, Chan AKH (2000) Dietary fibre intake and constipation in children with severe developmental disabilities. J Paediatr Child Health 36(3):236–239

Wong KH, Cheung PCK (2005) Dietary fiber from mushroom sclerotia. 1. Preparation, physico-chemical and functional properties. J Agric Food Chem 53:9395–9400

Wong KH, Cheung PCK (2009) Enzymatic preparation of mushroom dietary fiber: a comparison between analytical and industrial enzymes. Food Chem 115:795–800

Wu GD, Chen J, Hoffmann C et al (2011) Linking long-term dietary patterns with gut microbial enterotypes. Science 334(6052):105–108

Chapter 3
GRAS Fungi: A New Horizon in Safer Food Product

Nikita Singh and Smriti Gaur

3.1   Introduction

The ingredients added to our food items often go through an assessment test to be regarded as safe for consumption. Governmental organizations around the world make rules to regulate the same. Generally recognized as safe (GRAS) rules assigned to sections 201(s) and 409 of the Federal Food, Drug, and Cosmetic Act state that for the green light to be given by the US FDA, the ingredient which is added to the food item should go through an assessment test to be considered a GRAS element (Burdock and Carabin 2004).

The new rule called GRAS gives instructions to different food industries to label food additives as safe. The Food and Drug Administration (FDA) of the United States also allows experts, scientists, and researchers to determine any chemical, substance or additives added to food as GRAS, provided they can efficiently prove for the same. This facilitates exemption of food ingredients to go through the tough provisions which usually come under the Federal Food, Drug, and Cosmetic Act (FFDCA).The idea of food ingredients as "generally recognized as safe" was firstly represented by the Food Additives Amendment of 1958; afterwards all the introduced food additives were calculated by advanced measures (Goodrich 2010).

The perception of "safe use" in the food industry has risen in recent times in terms of safety determination, supervision and administration of food. "Safe use" recommends that the food which is being used for several generations by humans should not be harmful for consumption (Constable et al. 2007). It is necessary not only to show the importance of microorganism in the food product but also to certify that the microorganism which is present in food is beneficial or not for human consumption. Hence, it is important that no new microorganism or its by-product is

N. Singh · S. Gaur (✉)
Department of Biotechnology, Jaypee Institute of Information Technology, Noida, India
e-mail: smriti.gaur@jiit.ac.in

© The Author(s), under exclusive license to Springer Nature   27
Switzerland AG 2021
X. Dai et al. (eds.), *Fungi in Sustainable Food Production*, Fungal Biology,
https://doi.org/10.1007/978-3-030-64406-2_3

introduced in the market without proper assessment (Dufossé et al. 2005). The use of natural products in food enhances their health beneficial properties in an inexpensive and safe manner. Compounds produced by different fungal strains have different effects on human health. They can have varying activities depending not only on the age group but also on the health condition of an individual. Therefore, it becomes important to investigate them under different parameters (Kholer et al. 2015).

3.2 History of GRAS

The FDA rooted the Food Additives Amendment on January 1, 1958, and a GRAS list was prepared. It provides a checklist of more than 500 food ingredients that can be used as additives and hence were excluded from the confirmation procedure, and they demand for a pre-market calculation which is required for the assessment of food elements (Smith et al. 2001). Such food ingredients were defined as "generally recognized as safe" (GRAS) for their extensive use. However, any food constituent can be treated as "generally recognized as safe", if it is accepted by the experts and the scientists as safe for its comprehensive use for human consumption (Burdock and Carabin 2004).

There are two ways to consider any compound as GRAS, either by GRAS self-determination or by submitting the GRAS petition to the FDA. In GRAS self-determination, all the documents related to the ingredients are maintained and made accessible for purchase and administrative processes. On the other hand, in the GRAS petition, the FDA sanctions GRAS petition and the ingredients are indexed in the Code of Federal Regulations (CFR) (Frestedt 2018).

3.3 Fungi in All Quarters

Fungi exist anywhere in the environment. They can have very microscopic magnitude or may be present as fruiting bodies such as moulds or mushrooms. They have the ability to rapidly proliferate and can tolerate a large range of pH (Pitt and Hocking 1997; Perrone et al. 2007). Fungi live in a harmonious relationship with animals, plants, parasites and other fungi. They play a very important role in the environment as they have constitutional role in nutrient cycle and have the ability to dissolve biological material. They are directly used as food source such as mushrooms in human diet. They are also used in the distillation process for the preparation of food such as soya sauce, beer and wine. Fungi had been used in the manufacturing of antibiotics for a very long time, and more recently their role in the production of enzyme for use in detergent industries has also gained momentum. The advantageous trait of fungi is that they can be grown in ample dimension inclusive of extreme environment, whether it is beneath the sea (Rampelotto 2013), dessert (Vaupotic et al. 2008) or ionizing dispersal (Dadachova et al. 2007). Taxonomists have characterized more than 100,000 species of fungi but still a lot of research is needed to fully appreciate their global biodiversity (Mueller and Schmit 2006). Around 2.0–3.5 million species of fungi can be present on earth as estimated in the year 2017 (Hawksworth and Lucking 2001).

Fungi can be grown on soil due to the presence of a cluster of oxidative and hydrolytic enzymes in them which can disintegrate the lignocellulosic wall of plants. This feature of fungi facilitates for the decomposition of various biological substances like vegetables, fruits, herbs, beans, cereals, nuts, herbal drugs and woods (Baker 2006). Many fungi produce enzymes like pectinases, invertase, lactases, amylases and cellulases (Souza and Magalhães 2010). This property boosts the usefulness of fungi in the food and drug industry. However, not all fungi serve beneficial role to humans. Some fungi also show negative effects on human health. Healthy individuals can be infected by the protruding fungi, and other fungi can be invasive when the immune system is weak (Schardl et al. 2006). *Aspergillus niger* can cause serious health problems like lung infection, ear infection, abdominal pain and allergy in humans. Sometimes, it even weakens the immune system and causes immunotoxicity (Schuster et al. 2002). Some strains of *Aspergillus niger* cause carcinogenic effects on the liver and kidney. However, the harmful effects of fungi on human health can be reduced by appropriate classification of their different strains.

3.4 Fungi in Food

For the past few years, the use of fungi as an ingredient in food has been on the rise but they must be used with caution. Some fungi produce certain secondary metabolites such as mycotoxins which are harmful to humans. Earlier, *Aspergillus niger* was labelled as GRAS under the FDA which played a significant role in many food industries (Schuster et al. 2002). In contrast to its safe classification, it has been found that it is responsible for many human infections. It leads to respiratory failure, skin infection, weakening of the immune system, problems in the digestive tract, kidney and liver failure and nervous system problems (Schuster et al. 2002; May and Adams 1997). It is also linked with the decolourization and degradation of berry and vegetables, which leads to the lowering of their commercial value. It has been reported that mycotoxins have been found in some of the food products such as flour, cereals, legumes and some dried meat (Alshannaq and Yu 2017). Therefore, it has become necessary that before incorporating any fungi or its products in food, their safety and efficacy should properly assessed.

Fungi have been used in traditional food preparations for a long time. They are either consumed as whole or some of their inherent property is exploited for use in food preparations.

3.4.1   Directly Consumed as Food

Nutritious values of food can be supplemented by customizing the food with fungi. Fungi such as mushroom, add texture, flavour, nutritional and medicinal qualities to the food (Sanchez 2004). Amino acid, vitamins, minerals and lipids are some nutritive values which are provided by fungi. The reproductive fungal stage is generally called mushrooms. Mushrooms are plump, berry-bearing fruiting bodies of fungus. It has been seen that some of the species of mushroom grow overnight and expand on the surface of the ground. But actually, mushrooms take numerous days to grow and enlarge by the retention of fluid. *Parasola plicatilis* is one such mushroom which grows immediately, leaving no terrace on the next day after rainfall (Nagy et al. 2009), while *Pleurotus nebrodensis* may take many days to grow (Venturella 2006). The mycelia of fungi decay and recover organic material and take part in nutrient cycle. Many mushrooms are cultivated commercially in mushroom farms and then sold in the supermarket. Among them *Agaricus bisporus* is very famous among people, as it is grown in a guarded manner; hence, it is safe for human consumption. It acts as an important food source. These mushrooms are rich in water, carbohydrates, protein and fats too. Some mushrooms have a good amount of vitamin B such as niacin, and riboflavin, while some are a source of vitamin D (Simon et al. 2013).

3.4.2   Secondary Metabolites

Secondary metabolites are biological components produced by living organisms like plants, bacteria and fungi. Secondary metabolites are low molecular weight molecules. Unlike primary metabolites, they do not take part in reproduction, development and growth of organisms (Pusztahelyi et al. 2015). However, they have long-term effect on individual's stability. Fungi produce various secondary metabolites which help in the transcription of cells. Some of them have application as immunosuppressants and antibiotics (Brakhage 2013).

Terpenes, polyketides and non-ribosomal peptides are some of the important fungal metabolites. The first fungal secondary metabolite which was authorized by the FDA was lovastatin, which mainly helps in lowering the blood cholesterol level. The source of lovastatin is red yeast rice (Liu et al. 2006) and oyster mushrooms (Cimerman and Cimerman 1995).

3.4.3   Alcoholic Beverages

Alcoholic beverages are drinks produced by fermentation of sugar present in the grains, berries and fruits. In this process, liquid is distilled to an extremely alcoholic drink. In some parts of Nepal, Tibet and India, fermented by-products are widely used by the people in immense quantity. Some of these alcoholic beverages are prepared using fungi (Walker and Stewart 2016).

Kombucha, an alcoholic drink, referred as tea fungus or tea mushroom, is consumed by many people and has been shown to have numerous health benefits. This drink is green or black in colour. It is produced by symbiotic culture of bacteria and yeast (SCOBY). In this, the bacterial element is *Gluconacetobacter*, while the yeast element is *Saccharomyces cerevisiae* (Ayed et al. 2017). Upon consumption this drink can provide various health benefits such as anticancer activity, reduction in constipation, reduction in blood sugar level, etc. (Dufresne and Farnworth 2000).

3.5 Applications

Fungi are universal and are present in soil, water, air and organism or on the outer covering of organisms. They impact not only the life of a common man but also that of manufacturers, microbiologists and farm experts. They are not only used in preparation of food like soya sauce but also used in many industries to develop beneficial antibiotics and to command pest and microorganism outbreak. Hence, there are many applications of fungi.

3.5.1 Exploited as Nutraceuticals

Many food items are regarded as nutraceuticals. Nutraceuticals are those food substances which add health benefits and medical values to the consumers. However, many factors such as environmental pollution, high value, low production, resource misuse and inadequate raw material make their extraction very difficult (Liu et al. 2017). Many GRAS fungi have been identified for the preparation of nutraceuticals. Some nutraceutical productions have been accomplished by using strains such as *Sclerotium rolfsii* (Wang et al. 2016). Metabolic engineering through microbial production is also found to be an efficient and eco-friendly method for large-scale nutraceutical production. Commonly used fungi in food industries such as *Saccharomyces cerevisiae* can be metabolically regulated and can potentially produce polyamino acids, prebiotics and polysaccharides (Borodina and Nielson 2014).

3.5.2 Antioxidant Activity

Antioxidants are those compounds which inhibit oxidation. Oxidation is a chemical reaction which causes the degradation of food items. Hence, antioxidants stop hazardous chemical reaction caused by oxidants. Some of the antioxidants present in food items are selenium, vitamin E, vitamin C and carotenoids.

Some of the fungal strains such as *Lentinula edodes*, *Grifola frondosa*, *Pleurotus* spp., *Monascus purpureus* and *Trametes* are important sources of antioxidant (Helen et al. 2015). Another fungal strain known as *Salvia miltiorrhiza* is also a good source of antioxidant and is a Chinese vegetative drug (Chu-Yan et al. 2015). *Rhizopus oligosporus*, which has been regarded as GRAS, is used to increase the antioxidant activity. It has been studied that it increases the antioxidant activity of okara, a pulp which contains insoluble compound of soya bean (Sitanggang et al. 2019).

3.5.3   Increased Shelf-Life of Vegetables and Fruits

Spoilage of fruits and vegetables during post-harvesting has been a major concern. Chitosan is a linear polysaccharide which is composed of D-glucosamine and N-acetyl-D-glucosamine. It is declared as GRAS and has been used to protect fruits and vegetables after harvesting. It is mainly present in the outer layer of shrimps, crabs and fungi (Cheung et al. 2015). *Aspergillus niger is a very potent chitosan-producing fungus among different GRAS fungi (Logesh 2012)*. The chitosan layer of fruits and vegetables not only increases their shelf-life but also kills food-borne microorganisms. It helps in decreasing weight loss and the rate of respiration. Hence, fungal chitosan is found to be very vital in the shelf-life of fruits and vegetables (Romanazzi et al. 2012).

3.5.4   Flavouring Agent

There is a great demand for flavour and fragrance in the food industry. Nowadays, many synthetic food additives are being used as food flavouring agents. These additives not only add flavour to the food but also make it more appealing, but as they are synthetic, they can be the possible source of many diseases. Therefore, natural sources of flavours are always in high demand. There are some fungi which are found to be efficient flavour producers. *Aspergillus* and *Penicillium* are the species among fungi which produce rose and cheese flavours, respectively. *Penicillium rubrum* and *Penicillium purpurogenum* produce green pepper flavour when added to food items (Ali and Bibi 2010). Besides, yeast which is declared as GRAS is responsible for wine flavour. It also serves as a great source of nutrients such as amino acids and carbohydrates to various fermented foods (Raveendran et al. 2018).

3.5.5 Colouring Agent

Colour is suggested to play a crucial role in determining both the quality and acceptability of foods and beverages. Synthetic colour can have negative effects on the health of the consumers. Therefore, the demand for natural colour in food is always high. Plants and microorganisms are sources of natural colours. It has been studied and investigated that fungal species such as *Neurospora, Monascus, Fusarium* and *Penicillium* are great sources of natural pigments (Gmoser et al. 2017). Red pigments are extracted from *Monascus purpureus* and used as natural food colorants to many food items. These pigments not only add colour to the food but also provide health benefits to the consumers (Dufossé 2016).

3.6 List of GRAS-Certified Fungi and Their Products

The rich diversity of fungal species has been exploited by mankind for a very long period of time. They served as important sources of innovative food, extracellular enzymes or secondary metabolites owing to their peculiar growth conditions. Further, the ability of some of these fungi to secrete extracellular enzymes makes them viable commercial sources for extraction of these enzymes. Table 3.1 lists some of the GRAS-certified fungi for industrial production of certain fungal metabolites/enzymes.

Table 3.1 List of some of the GRAS-certified fungi for industrial production of certain fungal metabolites/enzymes

S.NO	Name of the fungi	Product obtained	References
1.	*Yarrowia lipolytica*	Citric acid	Carsanba et al. (2019)
2.	*Aspergillus fumigatus*	Amyloglucosidase enzyme	Pervez et al. (2019)
3.	*Rhizopus oryzae*	Carbohydrase enzyme	Prakasham et al. (2006)
4.	*Saccharomyces cerevisiae*	Baker's yeast protein	Teresa et al. (2017)
5.	*Fusarium moniliforme*	Gibberellic acid	Vidhya (2012)
6.	*Mucor miehei*	Esterase-lipase enzyme	Melani et al. (2019)
7.	*Aspergillus niger*	Cellulase and carbohydrase	Li et al. (2020)
8.	*Hericium erinaceum*	Milk protein enzymes	Sato et al. (2018)
9.	*Candida guilliermondii*	Citric acid	West (2016)
10.	*Saccharomyces fragilis*	Acid phosphatase and invertase enzyme	Davies (1953)

3.7   Concern of GRAS Ingredients

Despite the fact that food additives are regularly checked by regulatory agencies before their launch in the market, still some of the food industries add hundreds of food ingredients wilfully to human and animal foods without or very little knowledge by the government. That's where it makes a loophole for the food companies to escape from the diligent FDA legal inspection process (Manchanda et al. 2018).

There are so many food items in which food companies use ingredients such as aroma, chemical preservatives and/or colouring agents without consumer concern which can cause prompt danger to consumer health. It has been found that long-term usage of some of these food ingredients caused allergic reactions to consumers (Shaw et al. 1997).

There are some food ingredients which are classified as GRAS but are carcinogenic in nature and had been used in many food items for a very long time. Quorn, a readymade product and certified as GRAS, is a replacement of meat and is sold in more than 18 countries. It has been found to contain a mycoprotein as a main food ingredient which is a metabolic product of fungi *Fusarium venenatum* (Finnigan 2011). Quorn had faced many criticisms because of its bad health effects on the consumers. Still, there are many food additives present in the market which have been added in breakfast food, in frosted food, for stamina refreshment, etc. (Frank 2007).

Various food associations and organizations had shown their concerns towards the GRAS confirmation mechanism (Hallagan and Hall 2009). Many private organizations are also working on the improvisation of the GRAS confirmation process. Some strict regulations have to be set up to regulate the processing and regulation of additives, especially microorganisms such as fungi, in food industries to deal with any hazardous effects caused by them.

References

Ali M, Bibi A (2010) Production of pyrazine flavours by mycelial fungi (Doctoral dissertation, University of Pretoria)

Alshannaq A, Yu JH (2017) Occurrence, toxicity, and analysis of major mycotoxins in food. Int J Environ Res Public Health 14(6):632

Ayed L, Abid SB, Hamdi M (2017) Development of a beverage from red grape juice fermented with the Kombucha consortium. Ann Microbiol 67(1):111–121

Baker S (2006) *Aspergillus niger* genomics: past, present and into the future. Med Mycol 44:17–21

Borodina I, Nielsen J (2014) Advances in metabolic engineering of yeast *Saccharomyces cerevisiae* for production of chemicals. Biotechnol J 9(5):609–620

Brakhage A (2013) Regulation of fungal secondary metabolism. Nat Rev Microbiol 11:21–32

Burdock GA, Carabin IG (2004) Generally recognized as safe (GRAS): history and description. Toxicol Lett 150(1):3–18

Carsanba E, Papanikolaou S, Fickers P, Agriman B, Erten H (2019) Citric acid production by *Yarrowia lipolytica*. In: Sibirny A (ed) Non-conventional yeasts: from basic research to application, vol 38. Springer Nature, Cham, pp 91–117

Cheung RC, Ng TB, Wong JH, Chan WY (2015) Chitosan: an update on potential biomedical and pharmaceutical applications. Mar Drugs 13(8):5156–5186

Chun-Yan SU, Qian-Liang MI, Rahman K, Ting HA, Lu-Ping QI (2015) *Salvia miltiorrhiza*: traditional medicinal uses, chemistry, and pharmacology. Chin J Nat Med 13(3):163–182

Cimerman G, Cimerman A (1995) Pleurotus fruiting bodies contain the inhibitor of 3-hydroxy-3-methylglutaryl-coenzyme A reductase-lovastatin. Exp Mycol 19(1):1–6

Constable A, Jonas D, Cockburn A, Davi A, Edwards G, Hepburn P, Samuels F (2007) History of safe use as applied to the safety assessment of novel foods and foods derived from genetically modified organisms. Food Chem Toxicol 45(12):2513–2525

Dadachova E, Bryan RA, Huang X, Moadel T, Schweitzer AD, Aisen P, Nosanchuk JD, Casadevall A (2007) Ionizing radiation changes the electronic properties of melanin and enhances the growth of melanized fungi. PLoS One 2(5):457

Davies R (1953) Enzyme formation in Saccharomyces fragilis. 1. Invertase and raffinase. Biochem J 55(3):484

Dufossé L (2016) Current and potential natural pigments from microorganisms (bacteria, yeasts, fungi, microalgae). In: Handbook on natural pigments in food and beverages. Woodhead Publishing, Amsterdam, pp 337–354

Dufresne C, Farnworth E (2000) Tea, Kombucha, and health: a review. Food Res Int 33(6):409–421

Dufossé L, Galaup P, Yaron A, Arad SM, Blanc P, Murthy KNC, Ravishankar GA (2005) Microorganisms and microalgae as sources of pigments for food use: a scientific oddity or an industrial reality? Trend Food Sci Technol Elsevier 16(6):389–406

Finnigan TJA (2011) Mycoprotein: origins, production and properties. In Handbook of Food Proteins Woodhead Publishing, Cambridge, pp 335–352

Frank J (2007) Meat as a bad habit: a case for positive feedback in consumption preferences leading to lock-in. Rev Soc Econ 65(3):319–348

Frestedt JL (2018) Foods, food additives, and generally regarded as safe (GRAS) food assessments. In: Food control and biosecurity. Academic Press, London, pp 543–565

Goodrich, William W (2010) Address to the FFDCA concerning GRAS. Business Lawyer (Aba) heinonline.org 16:107

Gmoser R, Ferreira JA, Lennartsson PR, Taherzadeh MJ (2017) Filamentous ascomycetes fungi as a source of natural pigments. Fungal Biol & Biotechnol 4(1):4

Hawksworth DL, Lücking R (2001) Fungal diversity revisited: 2.2 to 3.8 million species. Microbiol Spectrum 105(12):1422–1432

Hallagan JB, Hall RL (2009) Under the conditions of intended use – new developments in the FEMA GRAS program and the safety assessment of flavor ingredients. Food Chem Toxicol 47(2):267–278

Helen S, Sean D, Richard M (2015) Filamentous fungi as source of natural antioxidant. Food Chem Elsivier 185:389–397

Köhler JR, Casadevall A, Perfect J (2015) The spectrum of fungi that infects humans. Cold Spring Harb Perspect Med 5(1):a019273

Liu L, Guan N, Li J, Shin HD, Du G, Chen J (2017) Development of GRAS strains for nutraceutical production using systems and synthetic biology approaches: advances and prospects. Crit Rev Biotechnol 37(2):139–150

Liu J, Zhang J, Shi Y, Grimsgaard S, Alraek T, Fønnebø V (2006) Chinese red yeast rice (*Monascus purpureus*) for primary hyperlipidemia: a meta-analysis of randomized controlled trials. Chinas Med 1(1):4

Li Q, Ray CS, Callow NV, Loman AA, Islam SM, Ju LK (2020) *Aspergillus niger* production of pectinase and α-galactosidase for enzymatic soy processing. Enzyme Microb Technol 134:109476

Logesh AR, Thillaimaharani KA, Sharmila K, Kalaiselvam M, Raffi SM (2012) Production of chitosan from endolichenic fungi isolated from mangrove environment and its antagonistic activity. Asian Pacific Journal of Tropical Biomedicine, 2(2):140–143

Manchanda S, Chandra A, Bandopadhyay S, Deb PK, Tekade RK (2018) Formulation additives used in pharmaceutical products: emphasis on regulatory perspectives and GRAS. In: Dosage form design considerations. Academic Press, London, pp 773–831

May GS, Adams TH (1997) The importance of fungi to man. Genome Res 7(11):1041–1044

Melani NB, Tambourgi EB, Silveira E (2019) Lipases: from production to applications. Sep & Purif Rev 12:1–6

Mueller GM, Schmit JP (2006) Fungal biodiversity: what do we know? What can we predict? Biodivers Conserv 16:1–5. https://doi.org/10.1007/s10531-006-9117-7

Nagy LG, Kocsubé S, Papp T, Vágvölgyi C (2009) Phylogeny and character evolution of the coprinoid mushroom genus *Parasola* as inferred from LSU and ITS nrDNA sequence data. Persoonia 22:28–37

Prakasham RS, Rao CS, Sarma PN (2006) Green gram husk – an inexpensive substrate for alkaline protease production by *Bacillus sp.* in solid-state fermentation. Bioresour Technol 97(13):1449–1454

Pitt JI, Hocking AD (1997) Fungi and food spoilage. Springer 10(8):593

Perrone G, Susca A, Cozzi G, Ehrlich K, Vargas J et al (2007) Biodiversity of Aspergillus species in some important agricultural products. Stud Mycol 59:53–66

Pervez S, Nawaz MA, Shahid F, Aman A, Tauseef I, Qader SAQ (2019) Characterization of cross-linked amyloglucosidase aggregates from *Aspergillus fumigatus* KIBGE-IB33 for continuous production of glucose. Int J Biol Macromol 135:1252–1260

Pusztahelyi T, Holb IJ, Pócsi I (2015) Secondary metabolites in fungus-plant interactions. Front Plant Sci 6:573

Rampelotto PH (2013) Extremophiles and Extreme Environments. Life MDPI 3(3):482–485

Raveendran S, Parameswaran B, Beevi US, Abraham A, Kuruvilla MA, Madhavan A, Pandey A (2018) Applications of microbial enzymes in food industry. Food Technol Biotechnol 56(1):16–30

Romanazzi G, Lichter A, Gabler FM, Smilanick JL (2012) Recent advances on the use of natural and safe alternatives to conventional methods to control postharvest gray mold of table grapes. Postharvest Biol Technol 63(1):141–147

Sánchez C (2004) Modern aspects of mushroom culture technology. Appl Microbiol Biotechnol 64:756–762

Sato K, Goto K, Suzuki A, Miura T, Endo M, Nakamura K, Tanimoto M (2018) Characterization of a milk-clotting enzyme from *Hericium erinaceum* and its proteolytic action on bovine caseins. Food Sci Technol Res 24(4):669–676

Schardl CL, Panaccione DG, Tudzynski P (2006) Ergot alkaloids – biology and molecular biology. Alkaloids Chem Biol 63:45–86

Schuster E, Dunn-Coleman N, Frisvad JC, Van Dijck PW (2002) On the safety of *Aspergillus niger* – a review. Appl Microbiol Biotechnol 59(4–5):426–435

Shaw D, Leon C, Kolev S, Murray V (1997) Traditional remedies and food supplements. Drug Saf 17(5):342–356

Simon RR, Borzelleca JF, DeLuca HF, Weaver CM (2013) Safety assessment of the post-harvest treatment of button mushrooms (*Agaricus bisporus*) using ultraviolet light. Food Chem Toxicol 56:278–289

Sitanggang AB, Sinaga WS, Wie F, Fernando F, Krusong W (2019) Enhanced antioxidant activity of okara through solid state fermentation of GRAS Fungi. Food Sci & Technol (AHEAD) 40(1):178

Smith RL, Doull J, Feron VJ, Goodman JI, Munro IC, Newberne PM, Portoghese PS, Waddell WJ, Wagner BM, Adams TB, McGowen MM (2001) GRAS flavoring substances 20 – the 20th publication by the expert panel of the flavor and extract manufacturers association on recent progress in the consideration of flavoring ingredients generally. Food Technol Chicago 55(12):34–55

Souza PM, Magalhães PO (2010) Application of microbial α-amylase in industry. A review. Braz J Microbiol 41(4):850–861

Teresa G, Vito P, Giorgio C, Giuseppa DB, Nicola C, Giacomo D (2017) Production of single cell protein (SCP) from food and agricultural waste by using *Saccharomyces cerevisiae*. Nat Prod Res 32(6):648–653

Vaupotic T, Veranic P, Jenoe P, Plemenitas A (2008) Mitochondrial mediation of environmental osmolytes discrimination during osmoadaptation in the extremely halotolerant black yeast Hortaea werneckii. Fungal Genet Biol 45(6):994–1007

Venturella G (2006) *Pleurotus nebrodensis*. In: IUCN. 2009. IUCN Red List of Threatened Species. Version 2009.1

Vidhya R (2012) Improved production of gibberellic acid by *Fusarium moniliforme*. J Microbiol Res 2(3):51–55

Walker GM, Stewart GG (2016) Saccharomyces cerevisiae in the production of fermented beverages. Beverages 2(4):1–30

Wang J, Guleria S, Koffas MA, Yan Y (2016) Microbial production of value-added nutraceuticals. Curr Opin Biotechnol 37:97–104

West TP (2016) A *Candida guilliermondii* lysine hyperproducer capable of elevated citric acid production. World J Microbiol Biotechnol 32(5):73

Chapter 4
Fungi in Food Bioprocessing

Rekha Kumari, Shashank Mishra, and Ashish Sachan

4.1 Introduction

For over a millennium, a majority of human population has taken advantage of natural fermentation processes to swage over a wide range of products including cheese, bread, yoghurt, beer, wine and different other beverages. Earlier the reasons behind these fermentative processes were not established, until relatively recently researchers tried to fill in the technical gap via exploring the world of microorganisms responsible for these processes.

Fermentation is often defined as a process where microbes act upon food considering it as an external source of energy. Fermentation basically converts sugars that are present in food into acids, thereby improving the taste, texture and odour of food products. This process also helps upon improving the food's storage life along with its quality and thus reduces the growth of unwanted microbes responsible for food spoilage. There are different microbes responsible for carrying out these fermentations. A major portion of fermented food products are produced using fungal species. Fungal fermented food products are those in which the fungal species (yeasts and moulds) primarily play an important in modifying the desirable qualities of fermented food products.

R. Kumari
Department of Bio-Engineering, Birla Institute of Technology, Mesra,
Ranchi, Jharkhand, India

S. Mishra (✉)
Quality Control and Quality Assurance Laboratory, Biotech Park,
Lucknow, Uttar Pradesh, India

A. Sachan (✉)
Department of Life Sciences, Central University of Jharkhand, Ranchi, Jharkhand, India
e-mail: ashish.sachan@cuj.ac.in

© The Author(s), under exclusive license to Springer Nature
Switzerland AG 2021
X. Dai et al. (eds.), *Fungi in Sustainable Food Production*, Fungal Biology,
https://doi.org/10.1007/978-3-030-64406-2_4

Some of the food products are prepared directly from microbes; certain others like mycoprotein, marketed as Quorn®, are manufactured from the mycelium of various cultured moulds, and another named as Marmite® is made from the yeast left over as byproduct from brewing. Nowadays edible mushrooms (the fruiting bodies of certain filamentous fungi) have become a part and parcel of our regular diets. Many of these fungal species are either harvested wild or cultivated. The most commonly grown mushroom in the UK which accounts for 95% of the total British market is *Agaricus bisporus*. There are several other fermented products like bread, cheese, tofu, truffles, tempe and koji, and certain beverages are produced industrially using fungal species as inoculant. Several sizes of mushrooms that are sold in the market correspond to different phases in the life cycle of mushroom. Certain exotic varieties like oyster mushrooms, shiitake and chanterelles are now sharing the supermarket shelves along with other edible mushrooms.

Microbes and fermentation, nowadays, are better understood and each process can be controlled mechanically to produce better quality products. Majority of fungal fermentation is carried out in Asian continent. The mycelium of fungi is often used by the virtue of their ability to degrade polymeric substances, as well as to improve upon the texture to a better extent.

4.2   Useful Fungi

There are several fungal species that we encounter in our daily lives which are relatively responsible for the production of fermented food products. These fungal species belong to different orders. Table 4.1 shows some of the major fungal species along with the fermented products they produce.

Enzymes have also been found to play an important role in food industry. There are different fungal species that produce certain enzymes which have paved their ways into food bioprocessing (Table 4.2).

4.3   Fermentation Processes

Since the fermented food products have different names in different regions, an attempt to classify these products on a regional basis is pretty worthy. Based on the kind of products formed, it will be easier to divide a line between beverages (alcoholic and non-alcoholic), condiments and flavouring agents (like soy sauces (Samson et al. 2004) and pastes), proteinaceous meat substitutes like tempe (Nout 1995) and bread or cake-like products (e.g. idli (Batra 1986)). The fermented products can also be classified based on the different fungal species involved in the fermentation process (Ko 1986; Nout et al. 2007). Basically there are three different types of fermentation processes (Nout and Aidoo 2002).

Table 4.1 Various fungal species used in food and beverage industry

Types of products	Fungal species	Product	Raw materials	Process	References
Asian foods	*Rhizopus oligosporus, R. oryzae* or *Mucor* sp.	Tempe	Soya beans	Multiple-stage fermentation	Ko (1986), Nout (1992) and Nout (1995)
	Monascus purpureus, M. anka or *M. ruber*	Red kojic rice, angkak	Rice	Natural fermentation	Lim (1991)
	Actinomucor elegans, A. taiwanensis, Mucor hiemalis, M. silvaticus, M. praini, M. subtillissimus or *Rhizopus chinensis*	Furu (sufu)	Soya beans	Solid-substrate fermentation	Wang and Fang (1986)
	Aspergillus oryzae or *A. sojae, Aspergillus tamarii* and *Saccharomyces rouxii*	Soy sauce	Soya beans, wheat	Solid-substrate fermentation	Ko (1986), Wang and Fang (1986), Nout (1995) and Fukushima (1998)
	Aspergillus oryzae, A. soyae, Saccharomyces rouxii, Candida etchellsii	Miso	Soya beans	Fermentation	Ko (1986), Wang and Fang (1986), Nout (1995) and Fukushima (1998)
	Fusarium venenatum	Quorn (mycoprotein)	Fungal mycelium	Fermentation	Hosseini et al. (2009)
	Tuber melanosporum	Truffles	Fruiting body	Fermentation	Talou et al. (1989) and Riccioni et al. (2008)
Bakery products	*Saccharomyces cerevisiae*	Bread	All-purpose flour	Solid-state fermentation	Mondal and Datta (2007)
	Saccharomyces cerevisiae	Bagels (Austrian bread)	All-purpose flour	Natural fermentation	Mondal and Datta (2007)
Cheese industry	*Penicillium candidum, P. camemberti*	Brie, Camembert and Limburger cheese	Milk	Solid-state fermentation	Hymery et al. (2014)
	Penicillium roqueforti	Roquefort cheese	Milk	Solid-state fermentation	Hymery et al. (2014)
Coffee	*Saccharomyces* sp.	Coffee powder	Coffee beans	Fermentation	Lee et al. (2015)

(continued)

Table 4.1 (continued)

Types of products	Fungal species	Product	Raw materials	Process	References
Alcoholic beverages	*Saccharomyces* sp./*Aspergillus oryzae*	Beer	Malt	Natural fermentation	Belitz et al. (2009)
		Vodka	Potatoes (starch)	Fermentation and distillation	Wisniewska et al. (2015)
		Tequila	Agave	Fermentation	Cedeno (2008)
		Wine	Fruit juice (grapes)	Natural fermentation	Mills et al. (2008)
		Rum	Molasses	Fermentation and distillation	Fahrasmane and Parfait (1998)
		Whiskey	Mashed grains	Fermentation and distillation	Walker and Hill (2016)
		Brandy	Fruit juice	Fermentation and distillation	Turk and Rozman (2002)
		Sake	Rice	Natural fermentation	Nout (1995) and Fukushima (1998)

Table 4.2 Fungal enzymes and their application

Fungal species	Enzyme	Applications
Penicillium notatum	Glucose oxidase	Used to remove oxygen from canned fruits, dried milk and other products
Fusarium moniliforme	Gibberellin	Plant hormone
Saccharomyces cerevisiae	Invertase	Used in preparation of soft centre candies, e.g. cordial cherries
Aspergillus sp.	Pectinase	Clarification of fruit juices
Aspergillus sp.	Amylase	Used for bread making and textile fibres
Aspergillus niger	₫-d-Galactosidase	Enzyme suppresses methane production in humans

4.3.1 Natural Fermentation

It is considered as one of the simplest types of fermentation which does not involve the addition of starter culture. The microorganisms responsible for such type of fermentation are present within the raw materials used in the process. In several cases the products are cooked after the fermentation process is over such as idli, nan, angkak (Lim 1991), etc.

4.3.2 Starter-Mediated Single-Stage Fermentation

In this type of fermentation, the raw materials are cooked prior to fermentation followed by the inoculation with the desired organisms. The organism upon incubation multiplies and ferments the food. The inoculant might be the spores or the mycelium of the fungus. The fermented food product is cooked before consumption such as kombucha (Nout 1992), airag (Naersong et al. 1996), etc.

4.3.3 Multiple-Stage Fermentation

In this process, two or more stages of fermentation are carried out. The organisms used in multiple stages can be either different or remain the same. One such fermentation is solid-substrate fermentation which is followed by liquid fermentation. In the first step the fungal organism produces certain enzymes to degrade high molecular weight polymers. The second step is the actual fermentation which is carried out using the enzyme produced by the microbe initially. There are several products which are produced using such kind of fermentation. For example, soy sauces (Fukushima 1998), beer (Nout 1992), wine (Fukushima 1998), etc.

4.4 Fermented Products: Manufacturing and Fungal Species Involved

4.4.1 Bread

Bread is one of the most commonly consumed dietary goods all around the world. Egyptians are considered as pioneers in the field of bread making and introducing this art to the world. The first production of bread took place in Neolithic era, some 12,000 years back (Mondal and Datta 2007). The most common ingredients used in bread are flour, water and yeast. Bread is a comparatively different food product than other commonly used baked goods as it is altogether a leavened product resulting from the fermentation of sugars present in wheat flour (Mondal and Datta 2007). Fresh bread usually gives a pleasant aroma and has a fine thin crust brown in colour which gives a soft feel to our taste buds (Giannou et al. 2003). With the advancement in technology and growing consumer demand for high-quality bread, several methods are developed in order to increase the shelf-life by adding certain additives such as emulsifiers and staling agents (Stampfli and Nersten 1995). Not only wheat, several other cereals (rye, millet, sorghum, maize, oats and millets) are used for the production of bread (Cauvain 2015). Though gluten is essential in bread making, researchers are trying to prepare gluten-free bread as some of the people across the world are allergic to gluten (Lamacchia et al. 2014). In order to replace gluten, certain additives are being used to compensate for its absence such as xanthan gum, guar gum, corn starch and eggs. These non-gluten components added in bread basically mimic the gluten network and thus help in improving the nutritional quality of bread (Mariotti et al. 2009).

4.4.1.1   Baker's Yeast

Commercially, baker's yeast is sold as a preparation of dried cells of one or more strains of fungus *Saccharomyces cerevisiae* (Ali et al. 2012). Yeast is used as a leavening agent in bread responsible for raising the bread. It is responsible for the fermentation of sugars present in wheat flour resulting in the production of carbon dioxide and ethanol as main products. The intensity of fermentation depends upon the form of yeast and the type of sugar present in dough (Hutkins 2006). In the process of bread making, the strains of *Saccharomyces cerevisiae* are subjected to several environmental and nutritional stresses such as high sucrose concentration, air drying and freeze-thawing (Attfield 1997). Sweet dough used for bread baking might contain 30% sucrose which can cause osmotic stress to the yeast, thereby damaging the cellular mechanism of yeast (Verstrepen et al. 2004). In order to minimize osmotic stress, yeast cells need to be first acclimatized to such higher sucrose concentration to carry out the fermentation (Tanaka et al. 2006).

There are several alternatives to baker's yeast as investigated by Jenson in 1998. One such yeast is *Kazachstania exigua*, previously known as *Saccharomyces exigua*, and the other which can be used is its anamorph *Candida holmii*. By the virtue of being highly osmotolerant and resistant to freezing, *Torulaspora delbrueckii* and its anamorph *Candida colliculosa* can also be used in dough fermentation (Gori et al. 2010). During dough fermentation, when the conditions become too inhibitory, most probably too acidic for *S. cerevisiae*, the microbial succession is often carried out using *Candida krusei* and its teleomorph *Issatchenkia orientalis* (Gori et al. 2010).

4.4.1.2   Baking Technology

Baking industry has evolved tremendously over the past 150 years. Earlier this art of baking was limited to small villages, but it paved its way to large-scale production in high technological industries. Several methods have evolved keeping in view the increasing demand of high-quality bread in the market (Decock and Cappelle 2005). Bakery goods are different from other products in which they are raised to produce goods of low density. In bakery goods, yeast is used as a leavening agent. Bakery goods are raised due to the production of carbon dioxide by fermentation of sugars. There are basically three methods by which bakery goods are produced (Mondal and Datta 2007). Straight dough method is the first method which is used in which mixing of all the ingredients is carried out in a single step. The ingredients of the dough might differ according to the availability of equipment and choice of manufacturer. Another method is the sponge and dough method in which the ingredients are mixed in two steps. In this, the leavening agent is prepared in the first step. The first step of this method involves the mixing of yeast, flour and water which is kept aside for several hours. The next step involves the mixing of other ingredients. The third method involves the mixing of all ingredients in one go in an ultrahigh mixer for few minutes. This method is commonly known as Chorleywood method (Giannou et al. 2003).

Bakers in large industries basically use two methods to prepare bread: mechanical dough development method and bulk fermentation method (O'Donnell 1996). The bulk fermentation method is a traditional method where the dough is kept untouched for few hours to ferment before moulding and baking. The basic steps involved in bread making are shown below in the form of a flowchart (Fig. 4.1).

It depends upon the manufacturer which method is being used for the production, but mixing of ingredients is an essential step as poor mixing of ingredients might result in the lower quality of the bread.

Bulk Fermentation Process

All the ingredients are mixed slowly to form an even mixture. The dough is then allowed to ferment slowly in a bowl for several hours. Earlier mixing results in the formation of a rough dough, which is just a dense mass of flour (non-extensible), but as soon as it is allowed to ferment, it forms a smooth surface and has gas bubbles trapped inside it. The dough is then finally allowed to rise and given the shape of a

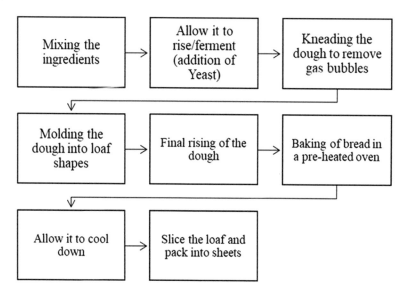

Fig. 4.1 Basic steps involved in bread making

loaf. It is allowed to be baked. There are four main factors that need to be kept in mind while using the bulk fermentation method (BRIT, Information Sheet 2010):

(a) Quantity of Yeast

Proof time will be affected by the size and the type of proofer. In ideal conditions, a minimum yeast level is preferred with maximum proof time. This also depends upon the type of bread being produced and the manufacturing process being used.

(b) Fermentation Time

Fermentation time usually depends upon the capabilities of plants in which bread is produced such as the number of bowls available for fermentation as well as the size of the dough room. With an increase in fermentation time, we can achieve maturity in dough at the divider.

(c) Dough Temperature

If the fermentation is carried out for longer duration, temperature should be comparatively lower. Ideally, dough should be at 26.5 °C at the divider. At higher temperature, yeast fermentation and reaction of oxidizing agent will be accelerated, thereby resulting in making the dough sticky in nature. Slow increase in dough temperature results in increasing the activity of yeast and oxidizing agent, thus affecting the maturity of dough.

(d) Oxidizing Agent

Oxidizing agents are also known as maturing agents. These are added in dough as they are responsible for strengthening the gluten structure and also provide dough the capacity to hold on to more carbon dioxide. This will result in proper rising of the loaf and even texture of bread. In order to adjust the maturity of the loaf, concentration of oxidizing agent needs to be optimized. Increase in the concentration of oxidizing agent will increase the maturity of dough.

The Mechanical Dough Development Method

In this method, dough formation takes place in shorter duration as all the ingredients are mixed at a higher speed and this energy is responsible for shearing the gluten structure and thus results in its reformation. With the use of this method, the dough rising time is reduced from 2 hours to 10 minutes due to development while mixing of ingredients.

4.4.2 Sufu (Furu)

Sufu, originally a Chinese fermented soya bean product, is highly flavoured and soft in nature like that of cheese and is creamy as well (Su 1986). Sufu is a rich source of protein commonly used as a side dish with breakfast (Han et al. 2001). It is not only produced in China but in other countries as well marketed under different names. In Japan, it is commonly known as *tofuyo*, *fu-nyu* or *nyu-fu* (Yasuda and Kobayashi 1989), *tao-hu-yi* in Thailand, *ta-huri* in the Philippines, *taokaoan* in Indonesia and *chao* in Vietnam (Beuchat 1995). Chinese people consider sufu a healthy food as it is easily digestible and a non-cholesterol food product from plant origin (Nout and Aidoo 2010). The variety of sufu is produced using different processes in different localities of China (Li 1997; Wang and Du 1998; Han et al. 2001).

Based on the technologies used for the production of sufu, there can be four different types of sufu whose base remains the same as that of curd obtained from soya bean by addition of calcium salts (Han et al. 2001).

4.4.2.1 Naturally Fermented Sufu

Four different steps are involved in the preparation of such sufu: (1) preparation of tofu, (2) fermentation of tofu to produce pizi, (3) salting and (4) ripening and processing.

4.4.2.2 Enzymatically Produced Sufu

Only three steps are involved in this process: (1) tofu preparation, (2) salting and (3) ripening. In order to carry out enzymatic fermentation before ripening, some amount of koji is added in dressing mixture to facilitate fermentation. This sort of ripening takes around 6–10 months.

4.4.2.3 Bacterial Fermented Sufu

Five steps are involved in such fermentation: (1) preparation of tofu, (2) pre-salting, (3) preparation of pizi by addition of bacterial culture, (4) salting and (5) ripening. In the pre-salting step, tofu is allowed to absorb salt in order to enhance the salt content up to 6.5%. Pure cultures of *Bacillus* sp. or *Micrococcus* sp. grown at 30–38 °C for about 1 week are used in the preparation of pehtze. The ripening time of this sufu is less than 3 months.

4.4.2.4 Fungal Fermented Sufu

Four steps are involved in this process: (1) preparation of tofu, (2) use of moulds and yeast to ferment tofu to produce pehtze, (3) salting and (4) ripening (discussed in later sections).

Based on different ingredients used in dressing mixture at ripening stage, sufu can be of different colours and flavours. The following four types of sufu have been identified in this category (Han et al. 2001).

4.4.2.5 Grey Sufu

The ripening of grey sufu is dominated by a combination of bacteria and fungi resulting in the production of strong and offensive odour in this sufu. The most common ingredients used in its dressing mixture are the soy whey left over from tofu, salts and some spices.

4.4.2.6 White Sufu

As compared to red sufu, white sufu is less salty and more famous in South China, but the ingredients used in its production are similar to that of red sufu except for the absence of red kojic rice. White sufu has an even distribution of light yellow colour both outside and inside.

4.4.2.7 Red Sufu

Red sufu is the most popular product in China because of its attractive colour and strong flavour. Red sufu shows a colour variation of red to purple from outside, and from inside, it has a colour range from light yellow to orange. The basic ingredients of red sufu comprise red kojic rice, alcoholic beverage, salt, sugar, spices and soya bean paste.

4.4.2.8 Other Types

There are several other ingredients added to the dressing mixture such as vegetables, rice and bacon and sometimes higher concentration of alcohol is preferred. Zui-Fang is the product resulting from dressing mixture containing higher levels of ethanol which is often termed as drunk sufu.

4.4.2.9 Mycology of Sufu

The fungal family involved in sufu formation is Mucoraceae. The most common genera involved in Sufu formation are *Actinomucor*, *Mucor* and *Rhizopus*. Some of the mycelial fungal species involved in furu production across China and neighbouring countries like Vietnam are *Mucor racemosus*, *Mucor circinelloides*, *Mucor hiemalis*, *Actinomucor repens*, *Rhizopus microsporus* var. *microsporus* and *Actinomucor taiwanensis* (Han et al. 2004). The mould used in pehtze formation is crucial and must possess certain characteristics to be an effective producer. Firstly, these moulds should have an efficient enzymatic system having high lipolytic and proteolytic activities as the raw substrate provided for pehtze formation is rich in lipid and protein and poor in carbohydrates. Secondly, the density of mycelial mat should be such that it should act as a protective layer for the final product of sufu. Another characteristic which these moulds should possess is the soothing colour of the mycelium in order to impart attractive colour to the product. Excess growth of these moulds during fermentation might result in off-odour, accumulation of mycotoxins and bacterial contamination which can adversely affect the quality of product.

The temperature that can be used for the growth of the above-mentioned mould responsible for sufu formation is 20–30 °C. According to the studies carried out by Han et al. in 2003, it was found that the optimum relative humidity required for pehtze formation was about 95–97%. The activity of β-glucosidase purified from *Actinomucor elegans* is affected by the change in incubation temperature of the mould (Yin et al. 2005). *Actinomucor elegans* and *Actinomucor taiwanensis* were found to be the best-suited moulds for pehtze development commercially in Beijing and Taiwan, respectively. Other popular starter cultures for furu development are *Mucor sufu* and *Mucor wutongqiao* (Han et al. 2004). Since these moulds grow well at 20–30 °C, it becomes really difficult to produce furu in hot summer. For the ease of furu production round the year, researchers screened several mutants produced using traditional mutagens and found they were capable of isolating *Mucor* sp. M_{263} and *Mucor* sp. H_4 surviving at 30–40 °C (Deng et al. 1996; Hu and Zhao 1998).

4.4.2.10 Manufacturing Process

Earlier, sufu was prepared in every household of China and Japan using the process of natural fermentation. This is still in practice in some parts of China such as Jiangsu and Zhejiang (Wang and Du 1998). Nowadays, furu is being produced at a large scale in industries following the same principle as that of traditional method. The basic steps involved in furu formation have been shown schematically in Fig. 4.2.

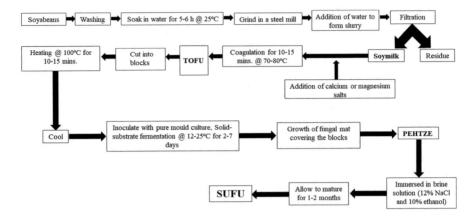

Fig. 4.2 Schematic diagram for industrial production of sufu (Adapted from Nout and Aidoo 2010)

Preparation of Tofu

This step is highly mechanized which involves less of human labour. There are several factors that affect the quality of tofu being produced such as soymilk characteristics (Lim et al. 1990), soya bean composition (Murphy et al. 1997), types of coagulants used (Moizuddin et al. 1999) and several other factors. High-quality soya bean is selected, washed and ground in a steel mill and converted into milk slurry by addition of water. The soymilk is then separated from the slurry by heating it and filtering it through a clean cloth. Next is the coagulation of soymilk by addition of calcium and magnesium sulphate at 70–80 °C. In order to obtain a homogenous coagulum, the mixture is agitated vigorously and then kept aside for 10–15 minutes to complete the coagulation process. Excess water is then removed by pressing the precipitate with cheese cloth in a mechanized press. Finally, the blocks of desired size are cut to obtain the final product (Nout and Aidoo 2010).

Preparation of Pehtze (Pizi)

The next stage involves fungal solid-substrate fermentation. The blocks of tofu are heated for 10–15 minutes at 100 °C to reduce the moisture content in the blocks to about 70%, thereby firming up the consistency of the tofu blocks and thus sterilizing the cubes before inoculating them with the pure fungal cultures (Nout and Aidoo 2010). Several fungal species such as *Mucor racemosus*, *Mucor circinelloides*, *Mucor hiemalis*, *Actinomucor repens*, *Rhizopus microspores* var. *microspores* and *Actinomucor taiwanensis* are used as starter culture. Depending upon the type of strain used, after 2–7 days of incubation at 12–25 °C, the mycelia of the culture cover the tofu blocks (Fukushima 1985). This product is now referred to as pehtze. For solid-substrate culture, the medium consists of bran and water (1:1.2–1.4), and for liquid substrate culture, maltose (2–3%) and peptone (1.5–2.0%) are added along with soy whey (Han et al. 2001).

Salting

Since the freshly prepared pehtze has no taste, salting and ripening are carried out to develop flavour and aroma in the pehtze. Pehtze is capable of absorbing salt and thus loses water until the salt content reaches up to a saturation level. This salt imparts certain taste to pehtze and thereby inhibits further fungal growth and also prevents the product from microbial contamination (Nout and Aidoo 2010). There are several ways in which salting is carried out. One such way involves transferring of pehtze to vessels containing saturated salt solution kept at room temperature. After 4–5 days, the salt content of pehtze reaches up to 12% and thereby reduces the moisture content up to 10–15%. In another method, pehtze is submerged in an alcoholic saline solution comprising 12% NaCl and 10% ethanol. This method is a combination of both salting and ripening in one step.

Ripening

In this stage, the enzymes produced by the mould act upon its respective substrates and thereby hydrolyze the protein and lipid, thus resulting in the production of principal compounds responsible for generating flavour and aroma in sufu. According to Hwan and Chou in 1999, the complicated flavour and aroma of sufu were reported to be the result of 22 esters, 18 alcohols, 7 ketones, 3 aldehydes, pyrazines, 3 phenols and other volatile compounds (Hwan and Chou 1999).

4.4.3 Red Kojic Rice

Red kojic rice, popularly termed as angkak, Chinese red rice, red qu or red yeast fermented rice, is conventionally obtained by fermentation of cooked rice using *Monascus* spp. such as *M. anka*, *M. purpureus* or *M. ruber*. Red fermented rice is referred to as koji in Japan, which has a specific aroma and purplish red tinge making it a potent colourant in red furu, red spirit and red rice (Chiao 1986). Several other researchers have also studied secondary metabolites being produced by *Monascus* species (Blanc et al. 1994; Lin et al. 2008).

4.4.3.1 Fungi Involved

In a liquid culture of *Monascus anka* MF107, the growth and pigment production were distinguished into three different phases (Fu et al. 1996). The first phase of liquid culture is dominated by mycelial growth and the pH of the media decreases from 5.5 to 4.6, and the second phase involves steady pigment production, thereby decreasing the pH from 4.6 to 8.4. And finally, the third phase is of gradual deterioration. *M. purpureus* is the most commonly used strain in red rice production. The mycelium of this strain is white in early stages. In due course of time, the mycelium changes to a rich pink colour and then to a distinctive orange yellow colour, thereby indicating the increased acidity of the medium. As the culture ages, a deep crimson colour is formed (INPR 2006).

Certain other strains of *Monascus* spp. are capable of producing mycotoxin citrinin. During fermentation, the levels of citrinin can be regulated to safe levels by the use of non-toxic strains and by optimizing culture conditions. However, uncontrolled starter cultures are used in many industrial operations while carrying out fermentation (Nout and Aidoo 2010).

4.4.3.2   Manufacturing Process

Since the texture of the product is solid, the method followed for its manufacturing is solid-substrate fermentation. Lucas et al. (1993) reported a swing bioreactor that provided mild agitation. The biggest advantage of using controlled reactors is that the environmental conditions can be controlled easily. Several optimum conditions for angkak production include 34% moisture content of the substrate and a fermentation period of 7 days at 28.8 °C.

First of all, the polished rice is soaked overnight in water, steamed and allowed to cool down. Then inoculation is carried out with the spores of *Monascus* spp. Incubation of about 1 to 2 weeks during solid-substrate fermentation allows the mould to grow and produce its secondary metabolites. Some of the major pigments produced include the purple pigment rubropunctamin, orange pigments rubropunctatin and monascorubrin and the yellow pigments monascin and ankaflavin (Pastrana et al. 1995). The schematic representation of angkak production has been shown in Fig. 4.3. These pigments are heat-stable and also capable of surviving in a wide range of pH. The nature and the quantity of pigment produced are purely dependent on the type of strain being used and the environmental conditions being produced for production (Miyake et al. 2008).

Several other cereals apart from rice are used as substrate for angkak production such as corn, but the high-quality product can only be achieved by optimizing the culture conditions. One such research suggested that addition of lactose and yeast extract in fermentation media resulted in high levels of pigment production (Pattanagul et al. 2007). According to Lee et al. (1995), solid-liquid fed-batch cultures gave a comparatively higher pigment yield. Different strains of *Monascus*, types of substrate, incubation temperature, pH and moisture content are some of the important factors indicating different production of their metabolites in angkak (Pattanagul et al. 2007).

Fig. 4.3 Traditional
manufacturing of red yeast
rice (Adapted from Nout
and Aidoo 2010)

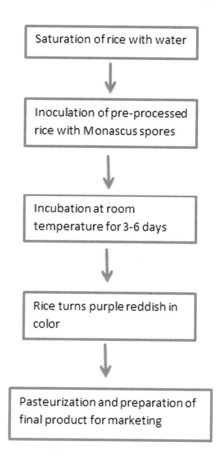

4.4.4   Soy Sauce

Soy sauce is a light to dark brown-coloured liquid often used in China as a condiment made from fermented paste of soya beans, brine and *Aspergillus oryzae* or *sojae* moulds (Leboffe and Pierce 2012). It has meat-like salty flavour which is commonly used in cooking (Nout and Aidoo 2010). Production of soy sauce is of ancient lineage, and today its production revolves around sophisticated levels of controlled production systems ensuring high-quality soy sauce. There are two specific fermentation processes involved in soy sauce production: one is koji fermentation with the use of *Aspergillus oryzae* or *Aspergillus sojae* strain and the other one is brine fermentation involving the use of lactic acid bacteria and yeast *Saccharomyces rouxii*.

4.4.4.1 Fungal Species Involved

There are several fungal species accepted for production of soy sauce such as *Aspergillus tamarii* (Raper and Fennell 1965), *Aspergillus sojae* (Murakami et al. 1982) and *Aspergillus oryzae* (Raper and Fennell 1965). These are often referred to as true koji moulds. These moulds can be differentiated based on their conidial head colour, size of conidiophores, vesicles, conidia and their growth pattern at 37 °C. A particular mould for koji production is selected based on certain factors such as its ability to sporulate, colour and flavour of the final product, length of stalk, enzyme production and its inability to produce toxins.

During koji fermentation, *A. oryzae* produces two main groups of enzymes: carbohydreases and proteinases (Aidoo et al. 1994; Chou and Rwan 1995). These groups of enzymes hydrolyze high molecular weight substrates into low molecular weight amino acids and peptides which are essential in brine fermentation step. The dominant moromi yeast in brine fermentation which grows to produce 3% alcohol is *Zygosaccharomyces rouxii*. There are several other compounds that add aroma to soy sauce which are produced by yeasts such as *Candida etchellsii* and *Candida versatilis* (Aidoo et al. 2006).

4.4.4.2 Manufacturing Process

A combined fermentation of soya beans, wheat grains, water and salt results in the production of soy sauce (Hesseltine 1977; Steinkraus 1989). The industrial production of fermented soy sauce involves three major steps: koji fermentation, brine fermentation and refining. A flowchart for the manufacturing of soy sauce is shown in Fig. 4.4.

Industrially, production of soy sauce at a large scale basically involves five operational units: the preparation of raw materials, koji process unit, brine fermentation unit, pressing and the final refining. Stacked wooden trays having a gap of about 10 cm are used in koji process, each of which is filled up to a depth of about 5–8 cm (Yokotsuka 1983; Aidoo et al. 1984). Traditionally the moromi process was carried out in ceramic pots or wooden tanks coated with resins. These have now been replaced by concrete or steel tanks having a capacity of 10–30 m^3.

Initially, the soya beans are soaked in water, boiled and mixed with the crushed or ground wheat. The moisture of this mixture needs to be maintained between 40% and 45%. This mixture is then placed in large wooden trays and mixed with mould *Aspergillus oryzae* or *Aspergillus sojae* and then allowed to ferment for about a week at 30 °C in order to form koji. The moisture content of koji should be about 25–30 °C. The function of mould here is to release hydrolytic enzymes to convert the substrate into low molecular weight compounds. In the next step, the koji is mixed with brine solution (23% w/v NaCl) in the ratio 1.0:1.5 to make moromi which later on undergoes fermentation for at least 1 year in order to develop colour and flavour in soy sauce. In the next step, the fermented moromi undergoes filtration. The soy sauce produced is boiled in order to sterilize it (Nout and Aidoo 2010).

Fig. 4.4 Schematic representation of soy sauce production (Adapted from Furusawa et al. 2013)

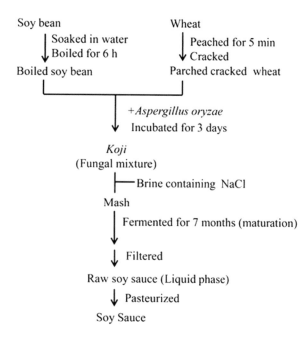

4.4.5   Tempe

Originated a hundred years ago, in Central and Eastern Java island (Shurtleff and Aoyagi 1979), tempe is considered as a staple food of Indonesia. Basically, tempe is a fermented product based on the solid state fermentation of soy beans involving the primary fungal strain of *Rhizopus* genus. There are several other ingredients apart from soy beans which can be used for tempe production. Tempe can be divided into five categories based on the raw materials used for its production: presscake residues, non-legume seeds, legumes, grains and a combination of both legumes and grains (Nout and Kiers 2005; Feng 2006). Since tempe is nutritionally rich and easy to digest, there is an increase in its demand by consumers who want to replace meat in their diet (Nout and Aidoo 2010).

4.4.5.1   Mycology of Tempe

Rhizopus microsporus is the major genus involved in tempe production, with several of its varieties such as *chinensis*, *microsporus*, *oligosporus* and *rhizopodiformis* (Liou et al. 1990; Nout and Rombouts 1990). An additional variety *tuberosus* involved in tempe production was described by Zheng and Chen 1990. The optimum temperature required for germination of sporangiospores in *R. oligosporus* is 37 °C and also that the accumulation of acetic acid is inhibitory (De Reu et al.

1995). It has been shown that *Rhizopus* sp. can survive under low oxygen concentration (0.2%), but not under absolute anaerobic conditions (Lin and Wang 1991).

It has also been seen that tempe-forming *Rhizopus* sp. were capable of utilizing raffinose as the sole carbon source for obtaining energy (Rehms and Barz 1995). Mitchell et al. (1990) described the growth of *Rhizopus oligosporus* sp. on a starchy model substrate as a multiple-stage process. The mycelium of the fungus penetrates up to 25% of the width of cotyledon by puncturing several layers of soya bean (Ko and Hesseltine 1979). An improved technique reported that tempe developed maximum strength after 30 h incubating at 30 °C, and later on the mycelium degenerated gradually (Ariffin et al. 1994). Mixed inoculation of *R. oryzae*, *Citrobacter freundii*, *R. oligosporus* and *Brevibacterium epidermidis* can result in the production of tempe nutritionally rich in niacin, tocopherol, pyridoxine, riboflavin, biotin, vitamin K and ergosterol (Wiesel et al. 1997).

4.4.5.2   Manufacturing Process

Predominantly, tempe is still a home-made or cottage-industry product manufactured in the areas of Indonesia and Malaysia. It is practically not known in Japan. It has recently been introduced to the USA and to Europe, specifically for the population of Indonesian origin and for the expanding Indonesian restaurant trade. It is made in modern, sanitary plants, using stainless steel equipment and sometimes, pure cultures of mould. The schematic representation of tempe production is shown in Fig. 4.5.

In small-scale industries in Indonesia, the dehulling of soya beans is carried out in a wet process. The major advantage of this process is that the beans suffer very less mechanical damage. This method is convenient only if sufficient water and cheap labour are available. The major disadvantage associated with dry dehulling in large-scale industries is the loss of soya beans. The split beans are then soaked in fresh water at 30 °C for 3–20 h in order to increase the moisture content of beans to enable microbial activity during fermentation. The soaking water is then discarded and the beans are cooked in fresh water. The cooking is carried out at approximately 100 °C for about 20–30 minutes. The cooking time varies depending upon the equipment being used. The hot water is then drained and the beans are exposed in air and spread out on trays to cool down. The beans are then inoculated with the sporangiospores *R. oryzae*, *R. oligosporus* and sometimes *Mucor* species. These are then mixed homogenously spread out in layered beds of 3–5 cm in thickness. Incubation of 2–3 days at 25–30 °C is sufficient enough for the spores to grow, thereby allowing the growth of mycelium to bind the beans together into sliceable cakes (Nout and Aidoo 2010).

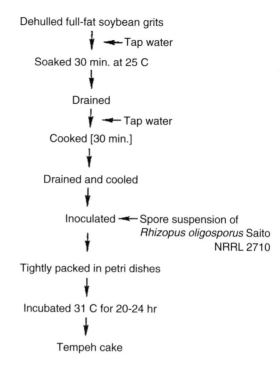

Fig. 4.5 Manufacturing process of tempe (Adapted from Wang 1984)

Dehulled full-fat soybean grits
↓ ◄— Tap water
Soaked 30 min. at 25 C
↓
Drained
↓ ◄— Tap water
Cooked [30 min.]
↓
Drained and cooled
↓
Inoculated ◄— Spore suspension of
Rhizopus oligosporus Saito
NRRL 2710
↓
Tightly packed in petri dishes
↓
Incubated 31 C for 20-24 hr
↓
Tempeh cake

4.4.6 Wine

Wine is one of the alcoholic beverages made from fermentation of grapes without the addition of any acid, enzymes, sugar, water or any other nutrient (Johnson 1989). The production of wine can date back to thousand years ago having early traces rooted to some parts of China (Hames 2010). According to our history, wine has always been consumed for its intoxicating effect. Wine holds religious importance as well. Rcd wine was thought to be associated with the blood of Egyptians. The most important organism responsible for wine fermentation is yeast. Yeast utilizes the sugar present in grapes and converts it into ethanol and carbon dioxide. Apart from grapes, there are several other substrates used to produce wine such as rice, plum, cherries, elderberry and pomegranate.

4.4.6.1 Mycology of Wine Fermentation

Saccharomyces cerevisiae is the major yeast responsible for wine fermentation. Vineyard is responsible for affecting the initial environment of wine fermentation (Mills et al. 2008). Microbes initially colonize around the stomata of grapes (Ribereau-Gaetyon et al. 2000). The most prevalent vineyard yeasts are *Hanseniaspora* and *Kloeckera* which are sexual anamorphs, representing over half of the yeast flora (Pretorius et al. 1999). Other yeast genera present on grapes include *Candida*, *Torulopsis*, *Zygosaccharomyces*, *Rhodotorula*, *Pichia*, *Metschnikowia* and *Cryptococcus* (Barnett et al. 1972; Rosini et al. 1982; Moore et al. 1988). There are several other yeasts present in the vineyard which affect wine fermentation: *Hansenula*, *Sporobolomyces* and *Kluyveromyces* (Davenport 1974).

Different species of yeast react differently towards different levels of ethanol. *S. cerevisiae* has the highest tolerance level towards ethanol among all the other yeasts involved in wine fermentation (Gao and Fleet 1988; Dittrich 1991). Temperature has also been found to affect ethanol tolerance among yeasts. Higher temperatures tend to decrease ethanol tolerance among yeasts. Some of the non-*Saccharomyces* yeasts present in the vineyard have also been found to produce ethanol: *Torulaspora delbrueckii* and *Saccharomycodes ludwigii* (Ciani and Maccarelli 1998). There are several other metabolites being produced by yeasts apart from ethanol. Other metabolites include octanoic and decanoic acids and medium-chain fatty acids which act by interfering with plasma integrity (Alexandre and Charpentier 1998; Bisson 1999).

4.4.6.2 Manufacturing Process

A schematic representation of wine manufacturing is shown in Fig. 4.6. Winemaking starts with the collection and crushing of grapes. For white wines, the grape juice can be separated away from the skins and clarified using any of the methods: cold settling, filtration or centrifugation. The juice is then transferred to a big barrel or fermentation tank and the alcoholic fermentation is carried out by yeasts primitive to the juice or by inoculation of a selected *S. cerevisiae* strain as starter culture. The fermentation is generally carried out for about 1 to 2 weeks at temperatures around 10–18 °C. Upon consumption of available sugars such as glucose and fructose present in grape juice, the wine is then considered "dry" and separated away from the yeast and grape lees (sediment).

Red wine production is slightly different than white wines. The crushed skins of grapes are left for fermentation to allow colour extraction. Like white wines, the alcoholic fermentation takes place either through the action of indigenous yeasts or by direct inoculation of a starter culture. During the fermentation the grape material tends to float to the top of the vat forming a "cap". To better enable extraction of red pigments and to influence wine flavour, winemakers typically punch down the cap or pump juice from the bottom over the cap. After a suitable period of time, the wine is separated from the grape skins and the fermentation is completed in another vessel. As described for white wines, the red wine is now "dry" and devoid of the main juice sugars.

Once the wine has been taken through the alcoholic and, if desired, the malolactic fermentation, it is often stored in tanks or barrels to allow flavour development. The residence time for storage is primarily determined by the style of wine and winemaker choice. Often white wines are not stored for long periods of time, while

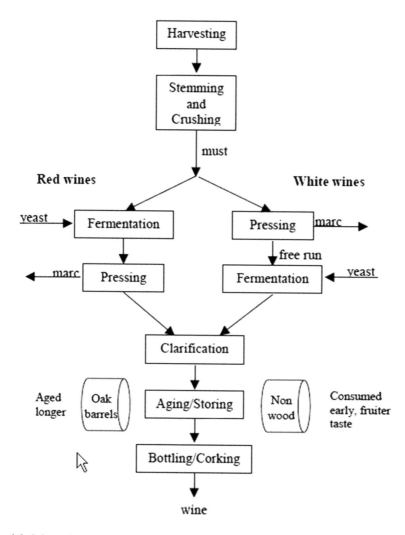

Fig. 4.6 Schematic representation of wine manufacturing (Adapted from Mills et al. 2008)

reds are frequently stored in oak barrels for several years. While the average wine contains approximately 13% ethanol, the alcohol by itself does not preclude future spoilage. Consequently winemakers must take great care to prevent exposure of the wine to oxygen, which can encourage microbial growth, as well as judiciously use antimicrobials (again, primarily sulphur dioxide) to prevent microbial spoilage (Mills et al. 2008).

4.5 Conclusion

The commercial exploitation of fungal species is very common these days as fungi serve as the best possible alternative both as a feed and food product. With the advancement in fermentation technology across the globe, it can be guaranteed that the quality of the products is being maintained to meet its consumer demands. Some of the traditional methods are still being practised at a small scale in villages. Some of the important fungal fermented food products have been discussed in this chapter. Biotechnologists are trying to continuously transfer the laboratory research to pilot scale which seems difficult in the case of fungi due to its complicated structure and growth pattern. The most common method used for fermentation in the case of fungi is solid-substrate fermentation, and the major disadvantage associated with this method is the heat and mass transfer. Researchers are continuously monitoring the other toxic metabolites being produced as byproducts of these fungal fermentations. Fungi being involved in fermentation processes have resulted in enhancing the food security and have also evolved as a promising food supplement for people. Extensive research is being carried out to isolate more novel fungal species which can be used in controlled systems to optimize process parameters to enhance the fungal production of food products. The need of the hour is to maximize productivity while minimizing health risks, and fungal species are acting as an excellent source of it.

References

Aidoo KE, Hendry R, Wood BJB (1984) Mechanized fermentation systems for the production of experimental soy sauce koji. J Food Technol 19:389–398

Aidoo KE, Smith JE, Wood BJB (1994) Industrial aspects of soy sauce fermentations using *Aspergillus*. In: Powell KA, Renwick A, Peberdy JF (eds) The genus *Aspergillus*: from taxonomy and genetics to industrial application. Springer, New York, pp 155–169

Aidoo KE, Nout MJR, Sarkar PK (2006) Occurrence and function of yeasts in Asian indigenous fermented foods. FEMS Yeast Res 6:30–39

Alexandre H, Charpentier C (1998) Biochemical aspects of stuck and sluggish fermentation in grape must. J Ind Microbiol Biotechnol 20:20–27

Ali A, Shehzad A, Khan MR, Shabbir MA, Amjid MR (2012) Yeast, its types and role in fermentation during bread making process – a review. Pak J Food Sci 22(3):171–179

Ariffin R, Apostolopoulos C, Graffham A, MacDougall D, Owens JD (1994) Assessment of hyphal binding in tempe. Lett Appl Microbiol 18:32–34

Attfield PV (1997) Stress tolerance: the key to effective strains of industrial baker's yeast. Nat Biotechnol 15:1351–1357

Barnett JA, Delaney MA, Jones E, Magson AB, Wich B (1972) The number of yeasts associated with wine grapes of Bordeaux. Arch Microbiol 83:52–55

Batra LR (1986) Microbiology of some fermented cereals and grain legumes of India and vicinity. In: Hesseltine CW, Wang HL (eds) . Carmer, Berlin, pp 85–104

Belitz HD, Grosch W, Schieberle P (2009) Food chemistry, 4th edn. Springer, Berlin

Beuchat LR (1995) Indigenous fermented foods. In: Reed G, Nagodawithana TW (eds) Enzymes, biomass, food and feed biotechnology. VCH Press, Weinheim, pp 523–525

Bisson LF (1999) Stuck and sluggish fermentations. Amer J Enol Vitic 50:107–119

Blanc PJ, Loret MO, Santerre AL, Pareilleux A (1994) Pigments of *Monascus*. J Food Sci 59:862–865

Breadmaking Information Sheet, The Science of Breadmaking. Baking Industry Research Trust. V1.0 (2010) (Available on- https://www.bakeinfo.co.nz/files/file/150/BIRT_Science_of_Breadmaking_Info_Sheet)

Cauvian S (2015) Technology of breadmaking, 3rd edn. Cham, Springer International

Cedeno MC (2008) Tequila production. Crit Rev Biotechnol 15(1):1–11

Chiao JS (1986) Modernization of traditional Chinese fermented foods and beverages. In: Hesseltine CW, Wang HL (eds) Indigenous fermented food of non – Western origin. Carmer, Berlin, pp 37–53

Chou CC, Rwan JH (1995) Mycelial propagation and enzyme production in koji prepared with *Aspergillus oryzae* on various rice extrudates and steamed rice. J Ferment Bioeng 79:509–512

Ciani M, Maccarelli F (1998) Oenological properties of non-*Saccharomyces* yeasts associated with winemaking. World J Microbiol Biotechnol 14:199–203

Davenport RR (1974) Microecology of yeasts and yeast-like organisms associated with an English vineyard. Vitis 13:123–130

De Reu JC, Rombouts FM, Nout MJR (1995) Influence of acidity and initial substrate temperature on germination of Rhizopus oligosporus sporangiospores in tempe fermentation. J Appl Bacteriol 78:200–208

Decock P, Cappelle S (2005) Bread technology and sourdough technology. Trends Food Sci Technol 16(1–3):113–120

Deng FX, Weng QY, Lu XC, Liu SC (1996) Screening of thermo-tolerant strain of *Mucor* sp. for sufu production. J China Brew Ind 1:27–31

Dittrich HH (1991) Wine and Branndy. In: Rehm HJ, Reed G (eds) Biotechnology: A multi-volume comprehensive treatise, Weinheim, New York, pp. 463–506

Fahrasmnae L, Parfait BG (1998) Microbial flora of rum fermentation media: a review. J Appl Microbiol 84:921–928

Feng X (2006) Microbial dynamics during barley tempeh fermentation (Doctoral thesis). Swedish University of Agricultural Sciences, Uppsala

Fu L, Zhou WB, Gao KR (1996) Fermentation characteristics of a high yielding strain of *Monascus anka*. Food Sci China 17:6–9

Fukushima D (1985) Fermented vegetable protein and related foods of Japan and China. Food Rev Int 1:149–209

Fukushima D (1998) Oriental fungal fermented foods. Int Mycol Congr Abstr 6:173

Furusawa R, Goto C, Satoh M, Nomi Y, Murata M (2013) Formation and distribution of 2,4-dihydroxy-2,5-dimethyl-3(2*H*)-thiophenone, a pigment, an aroma and a biologically active compound formed by the Maillard reaction, in foods and beverages. Food Funct 4:1076–1081

Gao C, Fleet GH (1988) The effects of temperature and pH on the ethanol tolerance of the wine yeasts *Saccharomyces cerevisiae, Candida stellate* and *Kloeckera apiculata*. J Appl Bacteriol 65:405–410

Giannou V, Kessoglou V, Tzia C (2003) Quality and safety characteristics of bread made from frozen dough. Trends Food Sci Technol 14(3):99–108

Gori K, Cantor MD, Jakobsen M, Jespersen L (2010) Production of bread, cheese and meat. In: Hofrichter M (ed) Industrial applications: the mycota. Springer-Verlag, Berlin, p 485

Hames G (2010) Alcohol in world history, 1st edn. Routledge, Abingdon

Han BZ, Rombouts FM, Nout MJR (2001) A Chinese fermented soybean food. Int J Food Microbiol 65:1–10

Han BZ, Kuijpers AFA, Thanh NV, Nout MJR (2004) Mucoraceous moulds involved in the commercial fermentation of Sufu Pehtze. Antonie Van Leeuwenhoek 85:253–257

Hesseltine CW (1977) Solid-state fermentation: process. Biochem 12:24–27

Hosseini SM, Khosravi KD, Mohammadifar MA, Nikoopour H (2009) Production of Mycoprotein by Fusarium venenatum growth on modified Vogel medium. Asian J Chem 21(5):4017–4022

Hu D, Zhao XH (1998) Physiology of thermo-tolerant strain of *Mucor* sp. H_4. J China Brew Ind 5:24–26

Hutkins RW (2006) Bread fermentation. In: Hutkins RW (ed) Microbiology and technology of fermented foods. Blackwell Publishing, Ames, pp 261–299

Hwan CH, Chou CC (1999) Volatile components of the Chinese fermented soya bean curd as affected by the addition of ethanol in ageing solution. J Sci Food Agric 79:243–248

Hymery N, Vasseur V, Coton M, Mounier J, Jany JL, Barbier G, Coton E et al (2014) Filamentous fungi and mycotoxins in cheese: a review. Compr Rev Food Sci Food Saf 13:437–456

INPR (The Institute for Natural Products Research). (2006). Cited in: http://www.jenshvass.com/pharmanex/pdf/inpr_monascus.pdf. (4/12/2006)

Johnson H (1989) Vintage: the story of wine, 1st edn. Simon and Schuster, New York

Ko SD (1986) Indonesian fermented foods not based on soybeans. In: Hesseltine CW, Wang HL (eds) . Carmer, Berlin, pp 67–84

Ko SD, Hesseltine CW (1979) Tempe and related foods. In: Rose AH (ed). Academic Press, London, pp 115–140

Lamacchia C, Camarca A, Picascia S, Diluccia A, Gianfrani C (2014) Cereal-based gluten-free food: how to reconcile nutritional and technological properties of wheat proteins with safety for celiac disease patients. Nutrients 6(2):575–590

Leboffe MJ, Pierce BE (2012) Microbiology laboratory theory and application, 3rd edn. Morton Publishing Company, Englewood

Lee YK, Chen DC, Chauvatcharin S, Seki T, Yoshida T (1995) Production of *Monascus* pigments by a solid liquid state culture method. J Ferment Bioeng 79:516–518

Lee LW, Cheong MW, Curran P, Yu B, Liu SQ (2015) Coffee fermentation and flavor – an intricate and delicate relationship. Food Chem 185:182–191

Li YJ (1997) Sufu – a health soybean food. J China Brew Ind 4:1–4

Lim G (1991) Indigenous fermented foods in South East Asia. ASEAN Food J 6:83–101

Lim BT, DeMan JM, DeMan L, Buzzell RI (1990) Yield and quality tofu as affected by soybean and soymilk characteristics, calcium sulfate coagulant. J Food Sci 55(4):1088–1111

Lin MS, Wang HH (1991) Anaerobic growth and oxygen toxicity of Rhizopus cultures isolated from starters made by solid state fermentation. Chin J Microbiol Immunol 24:229–239

Lin YL, Wang TH, Lee MH, Su NW (2008) Biologically active components and nutraceuticals in the *Monascus*-fermented rice: a review. Appl Microbiol Biotechnol 77:965–973

Liou GY, Chen CC, Chien CY, Hsu WH (1990) Atlas of the genus Rhizopus and its allies. Food Industry Research and Development Institute, Hsinchu

Lucas J, Schumacher J, Kunz B (1993) Solid-state fermentation of rice by *Monascus purpureus*. J Korean Soc Food Sci 9:149–159

Mariotti M, Lucisano M, Pagani MA, Ng PK (2009) The role of corn starch, amaranth flour, pea isolate, and *Psyllium* flour on the rheological properties and the ultrastructure of gluten free doughs. Food Res Int 42:963–975

Mills DA, Phister T, Neeley E, Johannsen E (2008) Wine fermentation. In: Cocolin L, Erocolini D (eds) Molecular techniques in the microbial ecology of fermented foods. Springer-Verlag, New York, pp 162–192

Mitchell DA, Greenfield PF, Doelle HW (1990) Mode of growth of *Rhizopus oligosporus* on a model substrate in solid state fermentation. World J Microbiol Biotechnol 6:201–208

Miyake T, Kong I, Nozaki N, Sammoto H (2008) Analysis of pigment compositions in various *Monascus* cultures. Food Sci Technol Res 14:194–197

Moizuddin S, Johnson LD, Wilson LA (1999) Rapid method for determining optimum coagulant concentration in tofu manufacture. J Food Sci 64(4):684–687

Mondal A, Datta AK (2007) Bread baking – a review. J Food Eng 86:465–474

Moore KJ, Johnson MG, Morris JR (1988) Indigenous yeast microflora on Arkansas white Riesling (*Vitis vinifera*) grapes and in model must systems. J Food Sci 53:1725–1728

Murakami H, Hayashi K, Ushijimi S (1982) Useful key characters separating three *Aspergillus* taxa – *A. sojae*, *A. parasiticus*, *A. toxicarius*. J Gen Appl Microbiol 28:55–60

Murphy PA, Chen HP, Hauck CC, Wilson LA (1997) Soybean protein composition and tofu quality. Food Technol 51(3):86–110

Naersong N, Tanaka Y, Mori N, Kitamoto Y (1996) Microbial flora of "airag", a traditional fermented milk of Inner-Mongolia in China. Anim Sci Technol 67:78–83

Nout MJR (1992) In: Carroll GC, Wicklow DT (eds) Ecological aspects of mixed-culture food fermentations. Dekker, New York, pp 817–851

Nout MJR (1995) Useful role of fungi in food processing. In: Samson RA, Hoekstra E, Frisvad JC, Filtenberg O (eds) . CBS, Utrecht, pp 295–303

Nout MJR, Aidoo KE (2002) Asian fungal fermented food. In: Osiewacz HD (ed) Industrial applications. The mycota. Springer-Verlag, Berlin, pp 23–47

Nout MJR, Aidoo KE (2010) Asian fungal fermented food. In: Hofrichter M (ed) . Springer-Verlag, Berlin Heidelberg, pp 29–58

Nout MJR, Kiers JL (2005) Tempe fermentation, innovation and functionality: update into the third millennium. J Appl Microbiol 98:789–805

Nout MJR, Rombouts FM (1990) Recent developments in tempe research. J Appl Bacteriol 69:609–633

Nout MJR, Sarkar PK, Beuchat LR (2007) In: Doyle MP, Beuchat LR (eds) Indigenous fermented foods. Springer Berlin Heidelberg, New York, pp 817–835

O'Donnell K (1996) Methods of bread dough making. AIB Res Depart Tech Bulletin 18(12)

Pastrana L, Blanc PJ, Santerre AL, Loret MO, Goma G (1995) Production of red pigments by *Monascus ruber* in synthetic media with a strictly controlled nitrogen source. Process Biochem 30:333–341

Pattanagul P, Pinthong R, Phianmongkhol A, Leksawasdi N (2007) Review of Angkak production (*Monascus purpureus*). Chiang Mai J Sci 34(3):319–328

Pretorius IS, Westhuizen TJ, Augustyn OPH (1999) Yeast biodiversity in vineyards and wineries and its importance to the South African wine industry: a review. S African J Enol Vitic 20:61–74

Raper KB, Fennell DI (1965) The genus Aspergillus. Williams and Wilkins, Baltimore

Rehms H, Barz W (1995) Degradation of stachyose, raffinose, melibiose and sucrose by different tempe-producing Rhizopus fungi. Appl Microbiol Biotechnol 44:47–52

Ribereau-Gaetyon P, Dubourdieu D, Doneche B, Lonvaud A (2000) Handbook of enology: the microbiology of wine and vinifications, 2nd edn. John Wiley and Sons, New York

Riccioni C, Belfiori B, Rubini A, Passeri V, Arcioni S, Paolocci F (2008) *Tuber melanosporum* outcrosses: analysis of the genetic diversity within and among its natural populations under this new scenario. New Phytol 180(2):466–478

Rosini G, Federici F, Martini A (1982) Yeast flora of grape berries during ripening. Microb Ecol 8:83–89

Samson RA, Hoekstra ES, Frisvad JC (2004) Introduction of food- and airborne fungi, 7th edn. CBS, Utrecht

Shurtleff W, Aoyagi A (1979) The book of tempeh. A super soyfood from Indonesia. Professional edition. Harper and Row, New York

Stampfli L, Nersten B (1995) Emulsifiers in bread making. Food Chem 52(4):353–360

Steinkraus KH (1989) Industrialization of indigenous fermented foods, 2nd edn. Marcel Dekker, New York

Su YC (1986) Sufu. In: Reddy NR, Pierson MD, Salunkhe DK (eds) Legume-based fermented foods. CRC, Boca Raton, pp 69–83

Talou T, Delmas M, Gaset A (1989) Analysis of headspace volatiles from entire black truffle (*Tuber melanosporum*). J Sci Food Agric 48:57–62

Tanaka F, Ando A, Nakamura T, Takagi H, Shima J (2006) Functional genomic analysis of commercial baker's yeast during initial stages of model dough-fermentation. Food Microbiol 23:717–728

Turk J, Rozman C (2002) A feasibility study of fruit brandy production. Agricultura 1:28–33

Verstrepen KJ, Iserentant D, Malcorps P, Derdelinckx G, Dijck VP, Winderickx J, Pretorius IS et al (2004) Glucose and sucrose: hazardous fast-food for industrial yeast? Trends Biotechnol 22(10):531–537

Walker GM, Hill AE (2016 Dec 20) Saccharomyces cerevisiae in the production of whisk(e)y. Beverages 2(4): 38. Available from: https://doi.org/10.3390/beverages2040038

Wang HL (1984) Tofu and tempeh as potential protein sources in the Western diet. J Amer Oil Chem Soc 61:528

Wang RZ, Du XX (1998) The production of Sufu in China. China Light Industry Press, Beijing

Wang HL, Fang SF (1986) History of Chinese fermented foods. In: Hesseltine CW, Wang HL (eds) Indigenous fermented food of non-western origin. Cramer, Berlin, pp 23–35

Wiesel I, Rehm HJ, Bisping B (1997) Improvement of tempe fermentations by application of mixed cultures consisting of *Rhizopus* sp. and bacterial strains. Appl Microbiol Biotechnol 47:218–225

Wisniewska P, Sliwinska M, Dymerski T, Wardencki W, Namiesnik J (2015) The analysis of vodka: a review paper. Food Anal Methods 8(8):2000–2010

Yasuda M, Kobayashi A (1989) Preparation and characterization of Tofuyo (fermented soybean curd). In: Ang HG, Nga ŽBH, Lim KK (eds) Proceedings of the 7th World Congress of Food Science and Technology, Singapore, pp. 82–86

Yin LJ, Li LT, Liu HE, Saito M, Tatsumi E (2005) Effects of fermentation temperature on the content and composition of isoflavones and B-glucosidase activity in sufu. Biosci Biotechnol Biochem 69:267–272

Yokotsuka T (1983) Scale up of traditional fermentation technology. Korean J Appl Microbiol Biotechnol 11:353–371

Zheng RY, Chen GQ (1990) *Rhizopus microsporus* var. *tuberosus* var. nov. Mycotaxon 69:181–186

Chapter 5
Fungal Productions of Biological Active Proteins

Gerardo Díaz-Godínez and Rubén Díaz

5.1 Fungi as Source of Bioactive Proteins

Proteins serve as nutrients for human and their quality is represented mainly by the amount of essential amino acid and their digestibility for absorption, but also, the nutraceutical effect has been observed in these molecules which promote health. It has been reported that most of the biological activities of proteins is given by certain peptide sequences which show their effect when exposed (Whisstock and Lesk 2003; Soriano-Santos et al. 2015). During food processing as well as during digestion (in vivo or by proteolytic enzymes), proteolysis occurs which generates peptides which may have biological or nutraceutical activity.

It is considered that any source of protein could present bioactive peptides, since they have been reported in proteins of animals, plants, fungi, and other microorganisms, and in many cases it has been observed that their effect or activity is inversely related to their size. Research to finding bioactive peptides has been carried out by hydrolyzing protein of food, but also bioinformatic studies have been carried out trying to predict the sequences with biological activity (Edwards et al. 2014; Soriano-Santos et al. 2015). The fermentative processes cause the hydrolysis of proteins generating peptides that could also have biological activity (García-Tejedor et al. 2014; Sanjukta and Rai 2016). Regardless of the technique used to obtain bioactive peptides, the next step is to assess whether to have biological activity using techniques in vitro or in vivo. In order to use bioactive peptides for therapeutic purposes, the cause and effect relationship between the consumption of bioactive peptides and their desired effects on human health should be known. The activities that have been reported in bioactive peptides are antihypertensive (with ribonuclease activity), antifungal, and antibacterial, among others (Daliri et al. 2017). The

G. Díaz-Godínez (✉) · R. Díaz
Research Center for Biological Sciences, Autonomous University of Tlaxcala,
Tlaxcala, Mexico

© The Author(s), under exclusive license to Springer Nature
Switzerland AG 2021
X. Dai et al. (eds.), *Fungi in Sustainable Food Production*, Fungal Biology,
https://doi.org/10.1007/978-3-030-64406-2_5

edible mushrooms, mainly basidiomycetes, have high nutritional value, promote the immune system, and as a source of natural antimicrobial substances have been used to cure bacterial infections since ancient times. Various kinds of proteins with several biological activities are produced by mushrooms. This review shows some bioactive peptides obtained from fungi (Table 5.1).

5.2 Inhibition of Angiotensin-Converting Enzyme by Mushroom

In this section, some works on peptides with inhibiting activity of angiotensin-I-converting enzyme (ACE) are reviewed. Hypertension is the main cause of disease in industrialized countries where 35% of mortality is due to this disease or some of its complications such as renal failure, heart disease, or cerebral hemorrhage (Mulero-Cánovas et al. 2011). ACE plays an important role in regulating blood pressure; within the renin-angiotensin system, ACE catalyzes the conversion of angiotensin I to angiotensin II, which is a compound with high vasoconstrictive potency, and its action causes the rapid contraction of the arterioles and, therefore, the increase in blood pressure, since it stimulates the secretion of aldosterone by the adrenal glands, a hormone that induces the excretion of potassium and the retention of sodium and water, causing the increase in extracellular volume, and neutralization of renin production. Renin releases decapeptide angiotensin I from the renin-angiotensin system. This system is, perhaps, the most important of the different vasoconstrictor and vasodilator mechanisms involved in the regulation of blood pressure, so by inhibiting the synthesis of angiotensin II, the increase in blood pressure is avoided (Mulero-Cánovas et al. 2011). Peptides that inhibit ACE have been found in proteins from different sources, with fungi being an alternative that has attracted the attention of many researchers.

In a recent study, the production of peptides with high inhibitory activity of angiotensin-converting enzyme (ACE) and low bitter taste was determined in 93 yeast strains isolated from Colombian kumis. By molecular identification it was found that *Galactomyces geotrichum*, *Pichia kudriavzevii*, *Clavispora lusitaniae*, and *Candida tropicalis* were the majority species. Eighteen strains produced fermented milk with ACE inhibitory activity (8.69–88.19%); however, the digestion of fermented milk samples by pepsin and pancreatin showed an increase in ACE inhibitory activity, and *Candida lusitaniae* KL4A was the best producer of peptides with ACE inhibitory activity. *Pichia kudriavzevii* KL84A and *Kluyveromyces marxianus* KL26A could be used as complementary primers in kumis, since they presented ACE inhibitory activity in fermented milk without bitter taste (Cháves-López et al. 2012).

Tricholoma matsutake is a highly prized mushroom. Recently, fruiting bodies were used to obtain an ACE inhibitory peptide with an IC_{50} of 0.40 μM. The peptide called TMP was purified and characterized from the crude water extract obtained

Table 5.1 Some bioactive peptides, source, and characteristics

Mushroom	Peptide or protein	Size (kDa)	Function	References
Polyporus adusta	Adustin	16.5	Translation-inhibiting polypeptide	Ng and Wang (2004)
Polyporus alveolaris	Alveolarin	14	Antifungal against *Botrytis cinerea, Fusarium oxysporum, Mycosphaerella arachidicola*, and *Physalospora piricola*	Wang et al. (2004)
Cordyceps militaris	Cordymin	10.906	Antifungal against *Bipolaris maydis, Mycosphaerella arachidicola, Rhizoctonia solani*, and *Candida albicans*	Wong et al. (2011)
			Inhibited HIV-1 reverse transcriptase, antiproliferative activity toward breast cancer cells	
Pleurotus eryngii	Eryngin	10	Antifungal against *Fusarium oxysporum* and *Mycosphaerella arachidicola*	Wang and Ng (2004b)
Ganoderma lucidum	Ganodermin	15	Antifungal against *Botrytis cinerea, Fusarium oxysporum*, and *Physalospora piricola*	Wang and Ng (2006)
Hypsizygus marmoreus	Hypsin	20	Antifungal against *Mycosphaerella arachidicola, Physalospora piricola, Fusarium oxysporum*, and *Botrytis cinerea*	Ng et al. (2010)
			Ribosome-inactivating protein (RIPs)	
			HIV-1 reverse transcriptase activity	
			Antiproliferative activity against mouse leukemia cells and human leukemia and hepatoma cells	

(continued)

Table 5.1 (continued)

Mushroom	Peptide or protein	Size (kDa)	Function	References
Flammulina velutipes	Velutin	13.8	Ribosome-inactivating protein (RIPs)	Wang and Ng (2001a, b, c), Ng et al. (2010)
			Inhibiting human immunodeficiency virus (HIV-1) reverse transcriptase, beta-glucosidase, and beta-glucuronidase	
Isaria felina	Isarfelin	NR	Antifungal against *Rhizoctonia solani* and *Sclerotinia sclerotiorum*	Guo et al. (2005a, b)
			Insecticidal activities from *Leucania separata*	
Lyophyllum shimeji	Lyophyllin	20	Ribosome-inactivating protein (RIPs)	Lam and Ng (2001b), Chan et al. (2010)
			Antifungal against *Physalospora piricola* and *Coprinus comatus*	
Hypsizygus marmoreus	Marmorin	0.9567	Ribosome inactivating protein (RIPs) with antiproliferative and HIV-1 reverse transcriptase inhibitory	Wong et al. (2008)
			Inhibited proliferation of hepatoma Hep G2 cells and breast cancer MCF-7 cells HIV-1 reverse transcriptase activity, and translation in the rabbit reticulocyte lysate system	
Lentinus squarrosulus Mont	Not named	17	Antifungal against *Trichophyton mentagrophytes*, *Trichophyton rubrum*, and *Candida tropicalis*	Poompouang and Suksomtip (2016)
Pleurotus ostreatus	Pleurostrin	7	Antifungal against *Fusarium oxysporum*, *Mycosphaerella arachidicola*, and *Physalospora piricola*	Chu et al. (2005)
Pleurotus tuber-regium	Pleuturegin	38	Ribosome-inactivating protein (RIP)	Wang and Ng (2001b)
			Inhibited translation in a cell-free rabbit reticulocyte lysate system	

(continued)

Table 5.1 (continued)

Mushroom	Peptide or protein	Size (kDa)	Function	References
Hypsizygus marmoreus	HM-af	9.5	Antifungal against *Flammulina velutipes*	Suzuki et al. (2011)
Lentinula edodes	Not named	87.2	Antimicrobial activity against *Escherichia coli* and *Staphylococcus aureus*	Sánchez-Minutti et al. (2016)
Termitomyces clypeatus, Pleurotus florida, Calocybe indica, Astraeus hygrometricus, Volvariella volvacea	Cibacron blue affinity eluted protein (CBAEP)	NR	Antiproliferative activity on several tumor cell lines through the induction of apoptosis Stimulatory effect on splenocytes, thymocytes, and bone marrow cells	Maiti et al. (2008)
			Immunostimulatory activity	
Clitocybe sinopica	Antibacterial protein (unnamed protein)	44	Antibacterial activity against *Agrobacterium rhizogenes, Agrobacterium tumefaciens, Agrobacterium vitis, Xanthomonas oryzae,* and *Xanthomonas malvacearum*	Zheng et al. (2010)
Flammulina velutipes	Fip-fve	NR	Fungal immunomodulatory proteins (FIPs)	Jiunn-Liang et al. (1995)
Ganoderma tsugae	Fip-gts	NR	Tumor immunotherapy	Wen-Huei et al. (1997)
Ganoderma lucidum	Ling zhi-8 (LZ-8)	NR	Immunomodulatory protein	Kino et al. (1989)
Hypsizygus marmoreus, Pleurotus ostreatus, Pleurotus sajor-caju, Russula delica, Tuber indicum	Ribonuclcases	NR	Ribonuclease activity	Ng et al. (2014)
Calvatia caelata	Ubiquitin-like peptide	NR	Translation-inhibiting protein and antiproliferative activity	Ng et al. (2014)
Russula paludosa	Peptide SU2	4.5	Antiviral properties	Wang et al. (2007)
Agrocybe cylindracea	Agrocybin	9	Antifungal activity	Ngai et al. (2005)

(continued)

Table 5.1 (continued)

Mushroom	Peptide or protein	Size (kDa)	Function	References
Cordyceps sinensis	Cordymin	10.906	Anti-inflammatory activity	Qian et al. (2011), Wang et al. (2012)
Cordyceps militaris				Wong et al. (2011)
Xylaria hypoxylon	Lectins	28.8	Antimitogenic and antitumor activities	Liu et al. (2006)
Oudemansiella canarii *Agaricus bisporus*	Not named	NR	Antifungal activity	Alves et al. (2013)
Agaricus bisporus	Not named	NR	Antibacterial activity against bacteria *Staphylococcus aureus*	Houshdar-Tehrani et al. (2012)
Tricholoma giganteum var. golden blessings	Trichogin	NR	Antifungal activity against *Fusarium oxysporum, Mycosphaerella arachidicola* y *Physalospora piricola*	Guo et al. (2005a, b)
			Inhibited HIV-1 reverse transcriptase	
Agaricus bisporus, Clitocybe nebularis, Cordyceps militaris, Flammulina velutipes, Ganoderma capense, Hericium erinaceum, Lactarius flavidulus, Pholiota adiposa, Pleurotus eous, Polyporus adusta, Russula delica, Rhipidura lepida, Schizophyllum commune, Tricholoma mongolicum, Xerocomus chrysenteron, Xylaria hypoxylon, Volvariella volvacea	Lectins	NR	Antiproliferative activity toward cancer cells in vitro and/or anticancer activity in vivo	Ng et al. (2014)

(continued)

Table 5.1 (continued)

Mushroom	Peptide or protein	Size (kDa)	Function	References
Pleurotus sajor-caju	Ribonuclease	12	Ribonuclease activity against *Fusarium oxysporum* and *Mycosphaerella arachidicola*	Ngai and Ng (2004)
			Antibacterial activity against *Pseudomonas aeruginosa* and *Staphylococcus aureus*	
			Antitumoral activity against cells HepG2 (hepatoma) and L1210 (leukemia)	
			Ribonuclease inhibited translation in a cell-free rabbit reticulocyte lysate system	

NR Not reported

from the fruiting bodies. LC-MS/MS analysis revealed its amino acid sequence (WALKGYK). The Lineweaver-Burk plots showed that TMP was a non-competitive inhibitor of ACE. TMP presented antihypertensive activity by a short-term assay (25 mg/kg reduced significantly the systolic blood pressure). TMP was very stable over a wide range of values of pH and temperatures. Other important effect that TMP demonstrated was the radical scavenging activity using 2,2-diphenyl-1-picrylhydrazyl (DPPH). All activities of TMP were corroborated by utilizing a synthetic peptide (Geng et al. 2016).

In another work, an water extract from mycelium of *Pleurotus pulmonarius* (gray oyster mushroom) had ACE inhibitory activity with IC_{50} value of 720 µg/ml and the purified protein increased 60 times the ACE inhibitory activity (IC_{50} = 12 µg/ml). Three proteins were responsible for the blood pressure lowering effects: serine proteinase inhibitor-like protein, nitrite reductase-like protein, and DEAD/DEAH box RNA helicase-like protein (Ibadallah et al. 2015). In a similar study, water extracts from fruiting body of *Pleurotus cornucopiae* showed antihypertensive effect on spontaneously hypertensive rats at a dosage of 600 mg/kg; after protein purification, two ACE inhibitory peptides with IC_{50} values of 0.46 and 1.14 mg/ml were obtained. The amino acid sequences of the two purified oligopeptides were RLPSEFDLSAFLRA and RLSGQTIEVTSEYLFRH. The molecular mass of the purified ACE inhibitors was 1622.85 and 2037.26 Da, respectively (Jeong-Hoon et al. 2011). The water extract (at 30 °C for 3 h) from fruiting body of *Tricholoma giganteum* was used for the extraction and characterization of an ACE inhibitory peptide. The extract and the purified peptide showed ACE inhibitory activities of IC_{50} = 0.31 mg and IC_{50} = 0.04 mg, respectively. The ACE inhibitory peptide was a novel tripeptide with very low similarity to other ACE inhibitors. The ACE activity was not affected after incubation with proteases. This ACE inhibitor had a clear antihypertensive effect in spontaneously hypertensive rats, at a dosage of 1 mg/kg (Lee et al. 2004).

ACE inhibitory activity (58.7%) was detected in a cold water extract of *Grifola frondosa* (IC_{50} of 0.95 mg). The extraction time increased the ACE inhibitory activity. After the purification of ACE inhibitory peptides was obtained an active fraction with an IC_{50} of 0.13 *mg* and a yield of 0.7%. The purified peptide showed competitive inhibition on ACE and maintained its inhibitory activity even after digestion by intestinal proteases (Choi et al. 2001).

A novel ACE inhibitory peptide (decapeptide) from *Saccharomyces cerevisiae* was purified and characterized. Maximal production of the ACE inhibitor was observed after 24 h of cultivation at 30 °C, and the activity was increased 1.5 times after treatment of the cell-free extract with pepsin. The purified peptide showed very low similarity to other ACE inhibitory peptide sequences, and its amino acid sequence was Tyr-Asp-Gly-Gly-Val-Phe-Arg-Val-Tyr-Thr. The purified inhibitor had a clear antihypertensive effect in spontaneously hypertensive rats (SHR) at a dosage of 1 mg/kg body weight (Jae-Ho et al. 2004).

The yeast *Malassezia pachydermatis* G-14 produced a novel ACE inhibitor when the strain was cultured in YEPD medium containing 0.5% yeast extract, 3.0% peptone, and 2.0% glucose at 30 °C for 24 h (Jeong et al. 2005).

A novel ACE inhibitory peptide (pentapeptide) was obtained from the water extract of the fruiting body of *Pholiota adiposa* ASI 24012. The purified ACE inhibitory peptide showed very low similarity to other ACE inhibitory peptide sequences. Its molecular mass was of 414 Da with a sequence of Gly-Glu-Gly-Gly-Pro and showed a clear antihypertensive effect on spontaneously hypertensive rats (SHR), at a dosage of 1 mg/kg (Kyo-Chul et al. 2006).

Recently, reishi (*Ganoderma lingzhi*) was studied to find ACE inhibitory peptides. To obtain auto-digested reishi (ADR) extract, reishi's protein was hydrolyzed by its own proteases. After 4 h administration of ADR to spontaneous hypertensive rats (SHRs), their systolic pressure significantly decreased to 34.3 mmHg (19.5% change) and the effect was observed up to 8 h of administration, with the decrease reaching as low as 26.8 mmHg. Eleven peptides were identified and four of them showed potent inhibition against ACE (IC_{50} values ranging from 73.1 μM to 162.7 μM) (Hai-Bang et al. 2014).

5.3 Antioxidant

Oxidative stress occurs when the production of harmful molecules, called free radicals (molecules highly unstable and reactive because they have one or more unpaired electrons), is greater than the protective capacity of antioxidant defenses. Within the radicals are the hydroxyl radical, superoxide anion radical, hydrogen peroxide, singlet oxygen, nitric oxide radical, and hypochlorite radical, among others; also smoking, ionizing radiation, pollution, organic solvents, and pesticides, among other factors, increase their production. The excess of free radicals attacks nucleic acids, proteins, enzymes, and other small molecules, causing them to lose their structure and function in the human body. There are many pathologies related to the presence of excess free radicals. Naturally there is a balance between the amount of free radicals that are produced in the body and the antioxidants that protect it; however, it is necessary to enrich the amount of antioxidants through the consumption of food to help the human body to reduce this oxidative damage. Currently, the scientific community works on the search for natural antioxidants (González-Palma et al. 2016).

Recently, yeasts were isolated from chhurpi with proteolytic activity to hydrolyze milk protein in search of the production of bioactive protein and enhance antioxidant property. A total of 125 proteolytic yeasts were isolated and molecular identification was made. *Kluyveromyces marxianus* and *Issatchenkia orientalis* were dominant in marketed products and *Trichosporon asahii, Saccharomyces cerevisiae*, and *Exophiala dermatitidis* from production centers. Fermentation using *Kluyveromyces marxianus* YMP45 and *Saccharomyces cerevisiae* YAM14 showed the higher antioxidant activity. The authors indicated that this study was the first on isolated yeast from fermented food of Northeast India for application in production of bioactive protein hydrolysate (Rai et al. 2016).

It has been reported that polysaccharides, polysaccharide-peptide complexes, and phenolic components are responsible for the antioxidant activity of the medicinal mushroom *Ganoderma lucidum*. However, research has shown that there is a peptide called *Ganoderma lucidum* (GLP) and that it is the main antioxidant component. GLP had potent antioxidant activity in both lightproof soybean oil and lard, tested by lipid peroxidation value. GLP showed a higher antioxidant activity in the soybean oil system, compared to butylated hydroxytoluene. GLP had scavenging activity toward hydroxyl radicals produced in a deoxyribose system with an IC_{50} value of 25 µg/ml. GLP inhibited the soybean lipoxygenase activity in a dose-dependent manner with an IC_{50} value of 27.1 µg/ml, and GLP effectively quenched superoxide radical anion produced by pyrogallol autoxidation in a dose-dependent manner. In rat liver tissue homogenates and in mitochondrial membrane peroxidation systems, a substantial antioxidant activity generated by GLP was observed. GLP in a dose-dependent manner acted blocking the auto-hemolysis of rat red blood cells (Sun et al. 2004).

Recently, the antioxidant activity of a peptide isolated from water extract of *Ganoderma lucidum* was evaluated. The molecular weight of purified peptide determined by SDS-PAGE, gel filtration chromatography, and MALDI-TOF MS was found to be 2.8, 3.34, and 3.35 kDa, respectively. This peptide was rich in phenylalanine, aspartic acid, proline, histidine, and isoleucine. It was suggested that the beneficial antioxidant properties in this mushroom may be due to low molecular weight and specific amino acid composition of the peptide (Girjal et al. 2012). A polypeptide with a molecular weight of 16.5 kDa was isolated from the mushroom *Polyporus adusta*. The polypeptide, called as adustin, inhibited translation in a cell-free rabbit reticulocyte lysate system with an IC_{50} of 0.34 µM (Ng and Wang 2004).

5.4　Antifungal

A polypeptide called alveolarin was isolated from fresh fruiting bodies of *Polyporus alveolaris* and showed antifungal activity against the mycelial growth of *Botrytis cinerea*, *Fusarium oxysporum*, *Mycosphaerella arachidicola*, and *Physalospora piricola*. Alveolarin had a molecular mass of 28 kDa (by gel filtration) but was observed in a single band with a molecular mass of 14 kDa in SDS-PAGE (Wang et al. 2004).

An antifungal peptide from fruiting bodies of the Thai common edible mushroom, *Lentinus squarrosulus* Mont., was obtained. Its molecular mass was about of 17 kDa by SDS-PAGE. The purified peptide did not show activity against both Gram-positive and Gram-negative bacteria, but it had a strong antifungal activity (comparable with ketoconazole) against several species of pathogenic fungi of human such as *Trichophyton mentagrophytes*, *Trichophyton rubrum*, and *Candida tropicalis* (Poompouang and Suksomtip 2016).

An antifungal protein designated trichogin was isolated from the mushroom *Tricholoma giganteum* var. golden blessings. Trichogin had antifungal activity against *Fusarium oxysporum*, *Mycosphaerella arachidicola*, and *Physalospora piricola*. It inhibited HIV-1 reverse transcriptase (IC_{50} of 83 nM) (Guo et al. 2005a, b).

A 7 kDa peptide isolated from fresh fruiting bodies of the oyster mushroom *Pleurotus ostreatus* showed inhibitory activity on mycelial growth of the fungi *Fusarium oxysporum*, *Mycosphaerella arachidicola*, and *Physalospora piricola*. The N-terminal sequence was different from known antifungal proteins and peptides (Chu et al. 2005).

A novel antifungal peptide was isolated from the aqueous extract of the medicinal mushroom *Cordyceps militaris* with a molecular mass of 10.906 kDa. The peptide was called cordymin and showed inhibition of mycelial growth of *Mycosphaerella arachidicola*, *Bipolaris maydis*, *Rhizoctonia solani*, and *Candida albicans* with an IC_{50} of 10 μM, 50 μM, 80 μM, and 0.75 mM, respectively; and there was no activity on *Aspergillus fumigatus*, *Fusarium oxysporum*, and *Valsa mali* with up to 2 mM. Cordymin also inhibited HIV-1 reverse transcriptase (IC_{50} of 55 μM) and presented antiproliferative activity toward breast cancer cells (MCF-7), while there was any effect on colon cancer cells (HT-29) (Wong et al. 2011).

A peptide designated as eryngin obtained from fruiting bodies of the mushroom *Pleurotus eryngii* inhibited the mycelial growth of *Fusarium oxysporum* and *Mycosphaerella arachidicola*. Eryngin showed a molecular mass of 10 kDa, and its N-terminal sequence was similar to a protein from the mushroom *Lyophyllum shimeji* with antifungal activity and little resemblance to thaumatin and thaumatin-like proteins (Wang and Ng 2004a).

A protein designated ganodermin with molecular mass of 15 kDa with antifungal activity was isolated from *Ganoderma lucidum* which is considered as medical mushroom. Ganodermin inhibited the mycelial growth of *Fusarium oxysporum*, *Botrytis cinerea*, and *Physalospora piricola* with an IC_{50} value of 12.4, 15.2, and

18.1 mM, respectively. It was devoid of hemagglutinating, deoxyribonuclease, ribonuclease, and protease inhibitory activities (Wang and Ng 2006).

A peptide called isarfelin was isolated from the mycelia of Isaria felina; this peptide showed inhibitory activity on mycelial growth in *Rhizoctonia solani* and *Sclerotinia sclerotiorum* and insecticidal activity against *Leucania separata*. However, it was devoid of activity toward several bacterial species including *Bacillus subtilis*, *Escherichia coli*, and *Staphylococcus aureus* (Guo et al. 2005a, b).

5.5 Ribosome-Inactivating Proteins

Ribosome-inactivating proteins (RIPs) are enzymes that act by inactivating ribosomes by removing one or more adenosine residues from rRNA. A novel RIP isolated from fruiting bodies of the mushroom *Hypsizygus marmoreus* with a molecular mass of 20 kDa was designated hypsin. This protein showed inhibitory activity against mycelial growth of *Mycosphaerella arachidicola*, *Physalospora piricola*, *Fusarium oxysporum*, and *Botrytis cinerea*, among others. Translation in the rabbit reticulocyte lysate system and HIV-1 reverse transcriptase activity was inhibited with IC_{50} of 7 nM and IC_{50} of 8 μM, respectively. Antiproliferative activity against mouse leukemia cells and human leukemia and hepatoma cells was observed. Only 40% of the translation-inhibitory activity was lost after heating at 100 °C for 10 min and no loss of this activity was observed after brief treatment with trypsin (Lam and Ng 2001a).

Hypsin from *Hypsizygus marmoreus* and velutin from *Flammulina velutipes* are two RIPs and their teratogenicity was evaluated using microinjection and postimplantation whole-embryo culture. Hypsin at 2.5 μM induced abnormal embryonic development during the organogenesis period from E8.5 to E9.5. There was an increase in the total number of abnormal embryos related to dosage increase, a drop in the final somite number, and a rise of abnormal structures. Open cranial neural tube, abnormal branchial arches, absence of forelimb buds, and twisted body axis were detected, while the otic and optic placodes were less affected. Histological study of the abnormal embryos revealed that the increase of cell death was correlated with abnormal structures, suggesting that induction of cell death by hypsin may account for its teratogenicity. Velutin did not exert any adverse influence on mouse development (Ng et al. 2010).

A novel RIP from the fruiting bodies of the mushroom *Lyophyllum shimeji* was isolated. The protein, designated lyophyllin, showed antifungal activity against *Physalospora piricola* and *Coprinus comatus*. Its molecular mass was 20 kDa and possessed an N-terminal sequence with some similarity to those of plant RIPs. It inhibited translation in rabbit reticulocyte lysate (IC_{50} of 1 nM), thymidine uptake by murine splenocytes (IC_{50} of 1 μM), and HIV-1 reverse transcriptase activity (IC_{50} of 7.9 nM). Lyophyllin did not manifest ribonuclease or hemagglutinating activity. In the same report, an antifungal protein called *Lyophyllum* antifungal protein (LAP) was first isolated from *Lyophyllum shimeji*. Its molecular mass was of 14 kDa and the N-terminal sequence similar to those of angiosperm thaumatin-like proteins and thaumatins and an inactive variant of the ubiquitin-conjugating enzyme. LAP showed antifungal activity against *Physalospora piricola* and *Mycosphaerella arachidicola*, whereas in *Rhizoctonia solani*, *Colletotrichum gossypii*, and *Coprinus comatus*, no effect was observed. Very low translation-inhibitory activity in a rabbit reticulocyte lysate system, insignificant ribonuclease activity toward yeast transfer RNA, and hemagglutinating activity toward rabbit erythrocytes were observed. HIV-1 reverse transcriptase was inhibited with an IC_{50} approximately of 5.2 nM. LAP and lyophyllin showed a synergism in antifungal activity against *Physalospora piricola* (Lam and Ng 2001b).

It has been reported that some plant RIPs adversely affect mouse embryonic development. Chan et al. (2010) reported for first the time on the teratogenicity of a mushroom RIP; this protein isolated from *Lyophyllum shimeji* was called lyophyllin. After it was partially sequenced, the translation-inhibitory activity was determined. Its teratogenicity was evaluated by using a technique entailing microinjection and postimplantation whole-embryo culture. Lyophyllin at a concentration as low as 50 µg/ml induced embryonic abnormalities during the period of organogenesis from E8.5 to E9.5. When the lyophyllin concentration was raised, the number of abnormal embryos increased, the final somite number decreased, and the abnormalities increased a lot. The affected embryonic structures was the cranial neural tube, forelimb buds, branchial arches, and body axis, while optic and otic placodes were less affected. Lyophyllin at a concentration higher than 500 µg/ml also induced forebrain blisters within the cranial mesenchyme. The histological study of the abnormal embryos reported an increase of cell death associated with abnormal structures indicating that cell death may be one of the underlying causes of teratogenicity of the mushroom RIP.

A RIP was purified from fresh sclerotia of the edible mushroom *Pleurotus tuberregium*. This RIP was designated pleuturegin and inhibited translation in a cell-free rabbit reticulocyte lysate system (IC_{50} of 0.5 nM). N-Terminal sequence of pleuturegin was different from RIPs flammulin and velutin from *Flammulina velutipes*, hypsin from *Hypsizygus marmoreus*, and lyophyllin from *Lyophyllum shimeji*, the only mushroom RIPs with known N-terminal sequences. The molecular weight of pleuturegin (38 kDa) was similar to that of flammulin (40 kDa) but larger than those of velutin, hypsin, and lyophyllin (13.8, 20, and 20 kDa, respectively). Pleuturegin lacked ribonuclease activity (Wang and Ng 2001a).

A novel RIP isolated from fresh fruiting bodies of the mushroom *Hypsizygus marmoreus* designated as marmorin reported a molecular mass of 9.567 kDa. Marmorin inhibited the proliferation of hepatoma Hep G2 cells (IC_{50} of 0.15 µM) and breast cancer MCF-7 cells (IC_{50} of 5 µM), as well as HIV-1 reverse transcriptase activity (IC_{50} of 30 µM) and translation in the rabbit reticulocyte lysate system (IC_{50} of 0.7 nM). Marmorin showed higher antiproliferative activity toward hepatoma (HepG2) and breast cancer (MCF-7) cells than RIPs from hairy gourd, bitter gourd, ridge gourd, garden pea, and the mushroom *Flammulina velutipes*, and similar inhibitory potency toward HIV-1 reverse transcriptase with those mentioned above with the exception of RIPS of ridge gourd and bitter gourd, because marmorin was more potent. It had less translation-inhibitory potency. The antifungal, protease, RNase, mitogenic, antimitogenic, nitric oxide-inducing, hemagglutinating, and trypsin inhibitory activities were not observed for marmorin (Wong et al. 2008).

5.6 Antibacterial

A protein purified from an aqueous extract of the dried fruiting bodies of *Lentinula edodes* showed growth inhibition of *Escherichia coli* and *Staphylococcus aureus*. The antimicrobial activity was evaluated by diffusion disk technique. The isolated protein (9.69 mg/l) formed bacterial inhibition zone of 2.48 cm for *Escherichia coli* and 2.68 cm for *Staphylococcus aureus* and their size was almost equal to that observed with 50 mg/l gentamicin. Its molecular weight was 87.2 kDa and has a specific activity of 28 cm/mg (Sánchez-Minutti et al. 2016).

A novel protein isolated from dried fruiting bodies of the wild mushroom *Clitocybe sinopica* showed antibacterial activity. The protein was composed of two subunits each with a molecular weight of 22 kDa (determined by SDS-PAGE), and its N-terminal amino acid sequence (SVQATVNGDKML) has not been reported for other antimicrobial proteins. The protein showed potent antibacterial activity against *Agrobacterium rhizogenes*, *Agrobacterium tumefaciens*, *Agrobacterium vitis*, *Xanthomonas oryzae*, and *Xanthomonas malvacearum* with a minimum inhibitory concentration, below 0.6 µM. However*, Pseudomonas batatae*, *Erwinia herbicola*, *Escherichia coli*, and *Staphylococcus aureus* were not affected. The antifungal activity assay against *Setosphaeria turcica*, *Fusarium oxysporum*, *Verticillium dahliae*, *Bipolaris maydis*, and *Bipolaris sativum* was negative. The antibacterial activity against *Agrobacterium tumefaciens* was stable after exposure to 20–60 °C for 30 min and to pH 4–9 for 1 h (Zheng et al. 2010).

The antibacterial activity of proteins isolated from fruiting bodies of *Agaricus bisporus* against both Gram-positive and Gram-negative bacteria was evaluated. The antibacterial activity of total extract proteins and protein fractions was evaluated by the method of microdilution. The isolated proteins showed growth inhibition against *Staphylococcus aureus* and methicillin-resistant *Staphylococcus aureus* (Houshdar-Tehrani et al. 2012).

A novel defensin called copsin was identified in the coprophilous basidiomycete *Coprinopsis cinerea*. The peptide was recombinantly produced in *Pichia pastoris*, and the three-dimensional structure was solved by NMR. Copsin was extremely stable against high temperatures and protease digestion possibly because the cysteine stabilized α-/β-fold with a unique disulfide connectivity, and an N-terminal pyroglutamate. Copsin had bactericidal activity against different Gram-positive bacteria, including human pathogens such as *Enterococcus faecium* and *Listeria monocytogenes*. Characterization of this activity revealed that copsin bound specifically to the peptidoglycan precursor lipid II and therefore interfered in the cell wall biosynthesis. In this case and unlike lantibiotics and other defensins, the third position of the lipid II pentapeptide is essential for effective copsin binding, which suggests that copsin serves as the basis for the synthesis of new antibiotics (Essig et al. 2014).

A ribonuclease (RNase) was isolated from fresh fruiting bodies of the mushroom *Pleurotus sajor-caju*. Its optimal pH and molecular mass were 5.5 and 12 kDa, respectively. It had preferential activity for poly (U) and with much lower activity toward poly (A), poly (G), and poly (C). The ribonuclease exhibited some activity toward herring sperm DNA and calf thymus DNA. Ribonuclease activity was not affected in the presence of KCl (10 and 100 mM) and NaCl (100 mM and 1 M), but was strongly inhibited by $CuSO_4$ (0.01 and 0.1 mM). Divalent salts such as $MgCl_2$, $CaCl_2$, $ZnCl_2$, $ZnSO_4$, and $FeSO_4$ inhibited with less power than $CuSO_4$. The ribonuclease was stable up to 60 °C for 1 h. The ribonuclease inhibited growth of bacteria *Pseudomonas aeruginosa* and *Staphylococcus aureus* and of the mycelial of the fungi *Fusarium oxysporum* and *Mycosphaerella arachidicola*. Ribonuclease decreased the viability of the tumor cells HepG2 (hepatoma) and L1210 (leukemia) which was reduced with an IC_{50} of 0.22 and 0.1 μM, respectively, and also inhibited translation in a cell-free rabbit reticulocyte lysate system (IC_{50} of 158 nM) and methyl-thymidine uptake by murine splenocytes (IC_{50} of 65 nM) (Ngai and Ng 2004).

An RNase (14.6 kDa) was isolated from dried fruiting bodies of the edible mushroom *Lactarius flavidulus*. Its optimum pH and temperature were 5.5 and 70 °C, respectively. The ranking of its activity toward various polyhomoribonucleotides was poly (C) > poly (G) > poly (A) > poly (U). It inhibited proliferation of HepG2 cells and L1210 cells with an IC_{50} of 3.19 μM and 6.52 μM, respectively. It also suppressed the activity of HIV-1 reverse transcriptase with an IC_{50} of 2.55 μM (Wu et al. 2012).

A novel 15 kDa RNase was purified from *Pleurotus djamor*. Its N-terminal amino acid sequence was different from those previously reported RNase sequences of mushrooms belonging to the genus of *Pleurotus* and other genera. It exhibited maximal RNase activity at pH 4.6 and 60 °C. Its activity toward polyhomoribonucleotides was poly (U) > poly (C) > poly (A) > poly (G). RNase inhibited proliferation of hepatoma cells and breast cancer cells. The ranking of inhibitory potencies of metal ions on RNase activity was Fe (3^+) > Al (3^+) > Ca (2^+) > Hg (2^+) (Wu et al. 2010).

An RNase was purified from dried fruiting bodies of the wild mushroom *Amanita hemibapha*. Its molecular mass was 45 kDa and exhibited maximal RNase activity at pH 5 and in a temperature range between 60 and 70 °C. It did not show ribonucleolytic activity toward four polyhomoribonucleotides. The amino acid sequence (GDDETFWEHEWAK) showed this RNase was a ribonuclease T2-like RNase. It had strong inhibitory activity against HIV-1 reverse transcriptase (HIV-1 RT) with an IC_{50} of 17 μM (Sekete et al. 2012).

A novel RNase with molecular mass of 18 kDa was purified from fresh fruiting bodies of *Hypsizygus marmoreus*. Its maximal RNase activity was at pH 5 and 70 °C. RNase showed the highest ribonucleolytic activity toward poly (C) (196 U/mg), followed by poly (A) (126 U/mg), and less activity toward poly (U) (48 U/mg) and poly (G) (41 U/mg). The RNase inhibited [^{3}H-methyl]-thymidine uptake by leukemia L1210 cells with an IC_{50} of 60 μM (Guan et al. 2007).

A peptide with RNase activity was isolated from fruiting bodies of the mushroom *Agrocybe cylindracea*. Its molecular mass was 9.5 kDa and demonstrating an N-terminal sequence similar to ubiquitin. It exhibited maximal RNase activity at pH 6 and the activity was stable over the temperature range of 0–60 °C. It showed antiproliferative activity on leukemia cell line (M1) and hepatoma cell line (HepG2) and also enhanced NO production in murine peritoneal macrophages with a potency comparable to that of lipopolysaccharide. RNase exerted activity preferentially on poly (C), much lower activity on poly (U), and negligible activity on poly (A) and poly (G) (Ngai et al. 2003).

A potent homodimeric ribonuclease exhibiting a molecular weight of 29 kDa was isolated from the fresh sclerotia of the mushroom *Pleurotus tuber-regium*. It showed strong ribonucleolytic activity toward poly (G), slight activity toward poly (U) and poly (A), and minimal activity toward poly (C). Its optimal pH was 6.5 using yeast transfer RNA as substrate. RNase activity was resistant to heating at 100 °C for 30 min but was inhibited by some salts. The protein inhibited cell-free translation in a rabbit reticulocyte lysate with an IC_{50} of 0.09 nM. Three of the four amino acid residues of the active site (positions 38–41) of the RNase of *Pleurotus ostreatus*, YNNF, were also found at positions 17–20 in the RNase of *Pleurotus tuber-regium*. However, unlike the RNase of *Pleurotus ostreatus*, no cysteine residues were detected in the N-terminal sequence (Wang and Ng 2001b).

A novel RNase has been purified from fruiting bodies of the mushroom *Russula virescens*. The RNase had a molecular mass of 28 kDa. In contrast to other mushroom RNases which are monospecific, it exhibited cospecificity toward poly (A) and poly (C). It showed an optimal pH of 4.5, which is lower than the values reported for other mushroom RNases, and an optimal temperature of 60 °C (Wang and Ng 2003a).

An RNase with a molecular mass of 28 kDa and specificity toward poly (U) and poly (A) and possessing an N-terminal sequence different to previously reported mushroom RNases was isolated from dried fruiting bodies of *Dictyophora indusiata* (veiled lady mushroom). It showed an RNase activity of 564 U/mg toward yeast transfer RNA. It demonstrated an optimal pH of 4–4.5 and an optimal temperature of 60 °C and at higher temperatures activity was lost (Wang and Ng 2003b).

The N-terminal sequence of an RNase purified from fruiting bodies of the silver plate mushroom *Clitocybe maxima* (family Tricholomataceae) exhibited some homology to ribonuclease from *Pleurotus ostreatus* (family Pleurotaceae). However, there is little similarity between the N-terminal sequences of RNases from various *Pleurotus* species, and less resemblance between RNases from *Clitocybe maxima* and *Pleurotus tuber-regium*. No structural relationship exists between RNases from *Clitocybe maxima* and those from *Volvariella volvacea*, *Lentinus edodes*, and *Irpex lacteus*. This RNase had a molecular mass of 17.5 kDa. It manifested approximately the same ribonucleolytic potency toward poly (A) and poly (G) followed by poly (U). Its activity toward poly (C) was comparatively very low. The optimum temperature and pH were 70 °C and 6.5–7.0, respectively (Wang and Ng 2004b).

An antiproliferative RNase with a new N-terminal sequence was isolated from fruiting bodies of the edible wild mushroom *Russula delica*. This novel RNase had a molecular mass of 14 kDa. Its optimum pH and temperature were pH 5 and 60 °C, respectively. The ranking of its activity toward various polyhomoribonucleotides was poly (C) > poly (G) > poly (A) > poly (U). This RNase could inhibit proliferation of HepG2 and MCF-7 cancer cells with an IC_{50} value of 8.6 μM and 7.2 μM, respectively. It did not contain antifungal activity or inhibitory activity of HIV-1 reverse transcriptase (Zhao et al. 2010).

An RNase (14.5 kDa) was isolated from fresh fruiting bodies of the edible mushroom *Lyophyllum shimeji*. Its optimum pH and temperature were pH 6 and 70 °C, respectively. It exhibited the highest ribonucleolytic potency toward poly (U), 25% as much activity toward poly (C), and no activity toward poly (A) and poly (G) was detectable. Its ribonucleolytic activity at 100 °C and 20 °C was similar. It inhibited proliferation of hepatoma HepG2 cells and breast cancer MCF7 cells with an IC_{50} of 10 μM and 6.2 μM, respectively and also inhibited the activity of HIV-1 reverse transcriptase with an IC_{50} of 7.2 μM (Zhang et al. 2010).

An RNase was purified from the fruiting bodies of the edible fungus *Pleurotus ostreatus*. Its N-terminal sequence was different from that of the ribonucleases of other fungi and the previously isolated *Pleurotus ostreatus* RNases. It exhibited a molecular mass of 12 kDa. The RNase displayed an activity of 11,490 U/mg on yeast tRNA. The highest ribonuclease activity was observed toward poly (U), followed by poly (A) and poly (C). No activity was shown toward poly (G). The optimum pH and temperature was 7 and 55 °C, respectively. It inhibited cell-free translation in a rabbit reticulocyte lysate with an IC_{50} of 240 nM (Ye and Ng 2003).

An RNase (29 kDa) cospecific for poly (A) and poly (U) was isolated from fruiting bodies of the mushroom *Boletus griseus*. Its N-terminal sequence showed some similarity to those of RNases from the mushrooms *Irpex lacteus* and *Lentinus edodes*. Its optimum temperature and pH were 60–70 °C and 3.5, respectively (Wang and Ng 2006).

A 20 kDa RNase was isolated from fresh fruiting bodies of the cultured mushroom *Schizophyllum commune*. It exhibited maximal RNase activity at pH 6.0 and 70 °C. Its highest ribonucleolytic activity was toward poly (U) (379.5 μ/mg), the second highest activity was toward poly (C) (244.7 μ/mg), and it had less activity toward poly (A) (167.4 μ/mg) and toward poly (G) (114.5 μ/mg). The RNase inhibited HIV-1 reverse transcriptase with an IC_{50} of 65 μM. Its N-terminal amino acid sequence was similar to that reported for *Volvariella volvacea* as well as in their optimum pH and polyhomoribonucleotide specificity, but some differences in chromatographic behavior and molecular mass were observed (Zhao et al. 2011).

5.7 Conclusions

Currently, chronic degenerative diseases are on the rise, so work must be done to change the eating habits and sedentary lifestyle. Mushrooms are an important source of bioactive peptides that should be included in the diet to help prevent and treat diseases.

References

Alves MJ, Ferreira ICFR, Días J, Teixeira V, Martins A, Pintado M (2013) A review on antifungal activity of mushroom (basidiomycetes) extracts and isolated compounds. Curr Top Med Chem 13:2648–2659

Chan WY, Ng TB, Lam JSY, Wong JH, Chu KT, Ngai PHK, Lam SK, Wang HW (2010) The mushroom ribosome-inactivating protein lyophyllin exerts deleterious effects on mouse embryonic development in vitro. Appl Microbiol Biotechnol 85(4):985–993

Cháves-López C, Tofalo R, Serio A, Paparella A, Sacchetti G, Suzzi G (2012) Yeasts from *Colombian kumis* as source of peptides with angiotensin I-converting enzyme (ACE) inhibitory activity in milk. Int J Food Microbiol 159:39–46

Choi HS, Cho HY, Yang HC, Ra KS, Suh HJ (2001) Angiotensin I-converting enzyme inhibitor from *Grifola frondosa*. Food Res Int 34(2–3):177–182

Chu KT, Xia L, Ng TB (2005) Pleurostrin, an antifungal peptide from the oyster mushroom. Peptides 26(11):2098–2103

Daliri EB, Oh DH, Lee BH (2017) Bioactive Peptides. Foods 6(5):32

Edwards RJ, Moran N, Devocelle M, Kiernan A, Meade G, Signac W, Foy M, Park SD, García-Tejedor A, Sánchez-Rivera L, Castelló-Ruiz M, Recio I, Salom JB, Manzanares P (2014) Novel antihypertensive lactoferrin-derived peptides produced by *Kluyveromyces marxianus*: gastrointestinal stability profile and in vivo angiotensin I-converting enzyme (ACE) inhibition. J Agric Food Chem 62(7):1609–1616

Essig A, Hofmann D, Münch D, Gayathri D, Künzler M, Kallio PT, Sahl HG, Wider G, Schneider T, Aebi M (2014) Copsin, a novel peptide-based fungal antibiotic interfering with the peptidoglycan synthesis. J Biol Chem 289(50):34953–34964

García-Tejedor A, Sánchez-Rivera L, Castelló-Ruiz M, Recio I, Salom JB, Manzanares P (2014) Novel antihypertensive lactoferrin-derived peptides produced by *Kluyveromyces marxianus*: gastrointestinal stability profile and *in vivo* angiotensin I-converting enzyme (ACE) inhibition. J Agric Food Chem 62(7):1609–1616

Geng X, Tian G, Zhang W, Zhao Y, Zhao L, Wang H, Ng TB (2016) A *Tricholoma matsutake* peptide with angiotensin converting enzyme inhibitory and antioxidative activities and antihypertensive effects in spontaneously hypertensive rats. Sci Rep 6:24130

Girjal VU, Neelagund S, Krishnappa M (2012) Antioxidant properties of the peptides isolated from *Ganoderma lucidum* fruiting body. Int J Pept Res Ther 18(4):319–325

González-Palma I, Escalona-Buendía HB, Ponce-Alquicira E, Téllez-Téllez M, Gupta VK, Díaz-Godínez G, Soriano-Santos (2016) Evaluation of the antioxidant activity of aqueous and methanol extracts of *Pleurotus ostreatus* in different growth stages. Front Microbiol 7:1099

Guan GP, Wang HX, Ng TB (2007) A novel ribonuclease with antiproliferative activity from fresh fruiting bodies of the edible mushroom *Hypsizigus marmoreus*. Biochim Biophys Acta 1770(12):1593–1597

Guo YX, Liu QH, Ng TB, Wang XH (2005a) Isarfelin, a peptide with antifungal and insecticidal activities from *Isaria felina*. Peptides 26(12):2384–2391

Guo Y, Wang H, Ng TB (2005b) Isolation of trichogin, an antifungal protein from fresh fruiting bodies of the edible mushroom *Tricholoma giganteum*. Peptides 26(4):575–580

Hai-Bang T, Yamamoto A, Matsumoto S, Ito H, Igami K, Miyazaki T, Kondo R, Shimizu K (2014) Hypotensive effects and angiotensin-converting enzyme inhibitory peptides of reishi (*Ganoderma lingzhi*) auto-digested extract. Molecules 19(9):13473–13485

Houshdar-Tehrani MH, Fakhrehoseini E, Nejad MK, Mehregan H, Hakemi-Vala M (2012) Search for proteins in the liquid extract of edible mushroom, *Agaricus bisporus*, and studying their antibacterial effects. Iran J Pharm Res 11(1):145–150

Ibadallah BX, Abdullah N, Shuib AS (2015) Identification of angiotensin-converting enzyme inhibitory proteins from mycelium of *Pleurotus pulmonarius* (oyster mushroom). Planta Med 81(2):123–129

Jae-Ho K, Dae-Hyoung L, Seoung-Chan J, Kun-Sub C, Jong-Soo L (2004) Characterization of antihypertensive angiotensin I-converting enzyme inhibitor from *Saccharomyces cerevisiae*. J Microbiol Biotechnol 14(6):1318–1323

Jeong SC, Kim JH, Kim NM, Lee JS (2005) Production of antihypertensive angiotensin I-converting enzyme inhibitor from *Malassezia pachydermatis* G-14. Mycobiology 33(3):142–146

Jeong-Hoon J, Seung-Chan J, Jeong-Han K, Yun-Hae L, Young-Cheoul J, Jong-Soo L (2011) Characterisation of a new antihypertensive angiotensin I-converting enzyme inhibitory peptide from *Pleurotus cornucopiae*. Food Chem 127(2):412–418

Jiunn-Liang K, Chyong-Ing H, Rong-Hwa L, Chuan-Liang K, Jung-Yaw L (1995) A new fungal immunomodulatory protein, FIP-fve isolated from the edible mushroom, *Flammulina velutipes* and its complete amino acid sequence. FEBS J 228(2):244–249

Kino K, Yamashita A, Yamaoka K, Watanabe J, Tanaka S, Ko K, Shimizu K, Tsunoo H (1989) Isolation and characterization of a new immunomodulatory protein, ling zhi-8 (LZ-8), from *Ganoderma lucidium*. J Biol Chem 264:472–478

Kyo-Chul K, Dae-Hyoung L, Jae-Ho K, Hyung-Eun Y, Jeong-Sik P, Jong-Soo L (2006) Production and characterization of antihypertensive angiotensin I-converting enzyme inhibitor from *Pholiota adiposa*. J Microbiol Biotechnol 16(5):757–763

Lam SK, Ng TB (2001a) Hypsin, a novel thermostable ribosome-inactivating protein with anti-fungal and antiproliferative activities from fruiting bodies of the edible mushroom *Hypsizigus marmoreus*. Biochem Biophys Res Commun 285(4):1071–1075

Lam SK, Ng TB (2001b) First simultaneous isolation of a ribosome inactivating protein and an antifungal protein from a mushroom (*Lyophyllum shimeji*) together with evidence for synergism of their antifungal effects. Arch Biochem Biophys 393(2):271–280

Lee HD, Kim HJ, Park SJ, Choi JY, Lee SJ (2004) Isolation and characterization of a novel angiotensin I-converting enzyme inhibitory peptide derived from the edible mushroom *Tricholoma giganteum*. Peptides 25(4):621–627

Liu QH, Wang HX, Ng TB (2006) First report of a xylose-specific lectin with potent hemag-glutinating, antiproliferative and anti-mitogenic activities from a wild ascomycete mushroom. Biochim Biophys Acta 1760(12):1914–1919

Maiti S, Bhutia SK, Mallick SK, Kumar A, Khadgi N, Maiti TK (2008) Antiproliferative and immunostimulatory protein fraction from edible mushrooms. Environ Toxicol Pharmacol 26(2):187–191

Mulero-Cánovas J, Zafrilla-Rentero P, Martínez-Cachá Martínez A, Leal-Hernández M, Abellán-Alemán J (2011) Péptidos bioactivos. Clin Invest Arterioscl 23(5):219–227

Ng TB, Wang H (2004) Adustin, a small translation-inhibiting polypeptide from fruiting bodies of the wild mushroom *Polyporus adusta*. Peptides 25(4):689–692

Ng TB, Lam JSY, Wong JH, Lam SK, Ngai PHK, Wang HX, Chu KT, Chan WY (2010) Differential abilities of the mushroom ribosome-inactivating proteins hypsin and velutin to perturb normal development of cultured mouse embryos. Toxicol In Vitro 24(4):1250–1257

Ng TB, Wong JH, Fai-Cheung RC, Tse TF, Tam T, Chan H (2014) Mushrooms: proteins, polysac-charidepeptide complexes and polysaccharides with antiproliferative and anticancer activities. Int J Cancer Res 7(3–4):287–300

Ngai PH, Ng TB (2004) A ribonuclease with antimicrobial, antimitogenic and antiproliferative activities from the edible mushroom *Pleurotus sajor-caju*. Peptides 25(1):11–17

Ngai PH, Wang HX, Ng TB (2003) Purification and characterization of a ubiquitin-like peptide with macrophage stimulating, antiproliferative and ribonuclease activities from the mushroom *Agrocybe cylindracea*. Peptides 24(5):639–645

Ngai PHK, Zhao Z, Ng TB (2005) Agrocybin, an antifungal peptide from the edible mushroom *Agrocybe cylindracea*. Peptides 26(2):191–196

Poompouang S, Suksomtip M (2016) Isolation and characterization of an antifungal peptide from fruiting bodies of edible mushroom *Lentinus squarrosulus* Mont. Malays J Microbiol 12(1):43–49

Qian GM, Pan GF, Guo JY (2011) Anti-inflammatory and antinociceptive effects of cordy-min, a peptide purified from the medicinal mushroom *Cordyceps sinensis*. Nat Prod Res 26(24):2358–2362

Rai AK, Kumari R, Sanjukta S, Sahoo D (2016) Production of bioactive protein hydrolysate using the yeasts isolated from soft *chhurpi*. Bioresour Technol 219:239–245

Sánchez-Minutti L, Téllez-Téllez M, Tlecuitl-Beristain S, Santos-López G, Díaz R, Kumar-Gupta V, Díaz-Godínez G (2016) Antimicrobial activity of a protein obtained from fruiting body of *Lentinula edodes* against *Escherichia coli* and *Staphylococcus aureus*. J Environ Biol 37:619–623

Sanjukta S, Rai AK (2016) Production of bioactive peptides during soybean fermentation and their potential health benefits. Trends Food Sci Technol 50:1–10

Sekete M, Ma D, Wang B, Wang H, Ng T (2012) First biochemical characterization of a novel ribonuclease from wild mushroom *Amanita hemibapha*. SpringerPlus 1(1):79

Soriano-Santos J, Reyes-Bautista R, Guerrero-Legarreta I, Ponce-Alquicira E, Escalona-Buendía HB, Almanza-Pérez JC, Díaz-Godínez G, Román-Ramos R (2015) Dipeptidyl peptidase IV inhibitory activity of protein hydrolyzates from *Amaranthus hypochondriacus* L. Grain and their influence on postprandial glycemia in streptozotocin-induced diabetic mice. Afr J Tradit Complement Altern Med 12(1):90–98

Sun J, He H, Xie BJ (2004) Novel antioxidant peptides from fermented mushroom *Ganoderma lucidum*. J Agric Food Chem 52(21):6646–6652

Suzuki T, Umehara K, Tashiro A, Kobayashi Y, Dohara H, Hirai H, Kawagishi H (2011) An anti-fungal protein from the culinary-medicinal beech mushroom, *Hypsizygus marmoreus* (Peck) Bigel. (Agaricomycetideae). Int J Med Mushr 13(1):27–31

Wang HX, Ng TB (2001a) Isolation of Pleuturegin, a novel ribosome-inactivating protein from fresh sclerotia of the edible mushroom *Pleurotus tuber-regium*. Biochem Biophys Res Commun 288(3):718–721

Wang HX, Ng TB (2001b) Purification and characterization of a potent homodimeric guanine-specific ribonuclease from fresh mushroom (*Pleurotus tuber-regium*) sclerotia. Int J Biochem Cell Biol 33(5):483–490

Wang H, Ng TB (2001c) Isolation and characterization of velutin, a novel low-molecular-weight ribosome-inactivating protein from winter mushroom (*Flammulina velutipes*) fruiting bodies. Life Sci 68(18):2151–2158

Wang H, Ng TB (2003a) A ribonuclease with distinctive features from the wild green-headed mushroom *Russulus virescens*. Biochem Biophys Res Commun 312(4):965–968

Wang H, Ng TB (2003b) A novel ribonuclease from the veiled lady mushroom *Dictyophora indusiata*. Biochem Cell Biol 81(6):373–377

Wang H, Ng TB (2004a) Isolation of a new ribonuclease from fruiting bodies of the silver plate mushroom *Clitocybe maxima*. Peptides 25(6):935–939

Wang H, Ng TB (2004b) Eryngin, a novel antifungal peptide from fruiting bodies of the edible mushroom *Pleurotus eryngii*. Peptides 25(1):1–5

Wang H, Ng TB (2006) Ganodermin, an antifungal protein from fruiting bodies of the medicinal mushroom *Ganoderma lucidum*. Peptides 27(1):27–30

Wang H, Ng TB, Liu Q (2004) Alveolarin, a novel antifungal polypeptide from the wild mushroom *Polyporus alveolaris*. Peptides 25(4):693–696

Wang JB, Wang HX, Ng TB (2007) A peptide with HIV-1 reverse transcriptase inhibitory activity from the medicinal mushroom *Russula paludosa*. Peptides 28(3):560–565

Wang J, Liu YM, Cao W et al (2012) Anti-inflammation and antioxidant effect of cordymin, a peptide purified from the medicinal mushroom *Cordyceps sinensis*, in middle cerebral artery occlusion-induced focal cerebral ischemia in rats. Metab Brain Dis 27(2):159–165

Wen-Huei L, Chih-Hung H, Chyong-Ing H, Jung-Yaw L (1997) Dimerization of the N-terminal amphipathic α-Helix domain of the fungal immunomodulatory protein from *Ganoderma tsugae* (Fip-gts) defined by a yeast two-hybrid system and site-directed mutagenesis. J Biol Chem 272:20044–20048

Whisstock JC, Lesk AM (2003) Prediction of protein function from protein sequence and structure. Q Rev Biophys 36(3):307–340

Wong JH, Wang HX, Ng TB (2008) Marmorin, a new ribosome inactivating protein with antiproliferative and HIV-1 reverse transcriptase inhibitory activities from the mushroom *Hypsizigus marmoreus*. Appl Microbiol Biotechnol 81(4):669–674

Wong JH, Ng TB, Wang H, Wing-Sze SC, Zhang KY, Li Q, Lu X (2011) Cordymin, an antifungal peptide from the medicinal fungus *Cordyceps militaris*. Phytomedicine 18(5):387–392

Wu X, Zheng S, Cui L, Wang H, Ng TB (2010) Isolation and characterization of a novel ribonuclease from the pink oyster mushroom *Pleurotus djamor*. J Gen Appl Microbiol 56(3):231–239

Wu Y, Wang H, Ng T (2012) Purification and characterization of a novel RNase with antiproliferative activity from the mushroom *Lactarius flavidulus*. J Antibiot 65:67–72

Ye XY, Ng TB (2003) Purification and characterization of a new ribonuclease from fruiting bodies of the oyster mushroom *Pleurotus ostreatus*. J Pept Sci 9(2):120–124

Zhang RY, Zhang GQ, Hu DD, Wang HX, Ng TB (2010) A novel ribonuclease with antiproliferative activity from fresh fruiting bodies of the edible mushroom *Lyophyllum shimeiji*. Biochem Genet 48(7–8):658–668

Zhao S, Zhao Y, Li S, Zhang G, Wang H, Ng TB (2010) An antiproliferative ribonuclease from fruiting bodies of the wild mushroom *Russula delica*. J Microbiol Biotechnol 20(4):693–699

Zhao YC, Zhang GQ, Ng TB, Wang HX (2011) A novel ribonuclease with potent HIV-1 reverse transcriptase inhibitory activity from cultured mushroom *Schizophyllum commune*. J Microbiol 49(5):803–808

Zheng S, Liu Q, Zhang G, Wang H, Ng TB (2010) Purification and characterization of an antibacterial protein from dried fruiting bodies of the wild mushroom *Clitocybe sinopica*. Acta Biochim Pol 57(1):43–48

Chapter 6
Fungal Pectinases: Production and Applications in Food Industries

Hamizah Suhaimi, Daniel Joe Dailin, Roslinda Abd Malek,
Siti Zulaiha Hanapi, Kugan Kumar Ambehabati, Ho Chin Keat,
Shanmuga Prakasham, Elsayed Ahmed Elsayed, Mailin Misson,
and Hesham El Enshasy

6.1 Introduction

Since 1968 until now, many studies had been done on pectinolytic enzyme or pectinase enzyme for commercial, eco-friendly, and applications in fruit or vegetable processing industries (Solis-Pereyra et al. 1993; Kashyap et al. 2001; Pedrolli et al. 2009). Nowadays, filamentous fungi are widely used in commercial enzyme pro-

H. Suhaimi · D. J. Dailin · R. A. Malek · S. Z. Hanapi · K. K. Ambehabati · H. C. Keat
S. Prakasham
Institute of Bioproduct Development (IBD), Universiti Teknologi Malaysia (UTM),
Johor Bahru, Malaysia

School of Chemical and Energy Engineering, Faculty of Engineering, Universiti Teknologi
Malaysia (UTM), Johor Bahru, Malaysia

E. A. Elsayed
Bioproducts Research Chair, Zoology Department, Faculty of Science, King Saud University,
Riyadh, Kingdom of Saudi Arabia

M. Misson
Biotechnology Research Institute, Universiti Malaysia Sabah, Jalan UMS,
Kota Kinabalu, Sabah, Malaysia

H. El Enshasy (✉)
Institute of Bioproduct Development (IBD), Universiti Teknologi Malaysia (UTM),
Johor Bahru, Malaysia

School of Chemical and Energy Engineering, Faculty of Engineering, Universiti Teknologi
Malaysia (UTM), Johor Bahru, Malaysia

Genetic Engineering and Biotechnology Research Institute, City of Scientific Research and
Technology Applications (CSAT), New Burg Al Arab Alexandria, Egypt
e-mail: henshasy@ibd.utm.my

© The Author(s), under exclusive license to Springer Nature
Switzerland AG 2021
X. Dai et al. (eds.), *Fungi in Sustainable Food Production*, Fungal Biology,
https://doi.org/10.1007/978-3-030-64406-2_6

duction in food and beverage industries especially those that belong to *Aspergillus* family. This is due to their high capacity for enzyme production and their safety category as GRAS (generally recognized as safe) microbes according to the FDA (Food and Drug Administration, USA) (Schuster et al. 2002; Iwashita 2002; Leyva et al. 2017). In this regard, many studies are focusing on the efficient production and secretion of pectinase particularly from *Aspergillus niger* (Sarrouh et al. 2012; Poletto et al. 2015; Suhaimi et al. 2016; El-Enshasy et al. 2018).

Pectinase enzyme is widely used in different applications with big market in food processing industries (Li et al. 2012). In 2015, the market of technical enzyme was estimated to increase at 6.6% compound annual growth rate (CAGR) to reach $1.5 billion with the highest sales in the leather market and bioethanol market (Sarrouh et al. 2012). The pectinase enzyme is a fundamental piece of the nourishment industry. The assessed sale estimation of every single enzyme in 1995 was 1 billion dollar, of which 75 million dollars was for pectinases. In the current biotechnological time, pectinase is one of the approaching chemicals demonstrating dynamic increment in their market. They kept up the normal yearly development pace of 2.86% from USD 27.6 million in 2013 to USD 30 million in 2016, and it is assessed that by 2021, the market size of the pectinase will reach USD 35.5 million (Global Pectinase Market Research Report 2017; Kavuthodi and Sebastian 2018).

The technical application of pectinase in textile industry is to destabilize the outer cell to improve fiber extraction. Furthermore, pectinase was used in food processing especially in the juice industry for degrading the pectin which is structural polysaccharide present in the cell wall and to increase the overall juice production. However, about 75% of pectinases in the global market are manufactured by three world's huge enzyme producer companies, including Denmark-based Novozymes, US-based DuPont (through May 2011 acquisition of Denmark-based Danisco) and Switzerland-based Roche (Li et al. 2012; Sarrouh et al. 2012).

Microbial production of pectinase includes mainly yeast and mold. Among the genera sources, the most typically studied in the last 15 years were *Erwinia*, *Bacillus*, *Saccharomyces*, *Kluyveromyces*, *Aspergillus*, *Penicillium*, *Fusarium*, and *Rhizopus* (Favela-Torres et al. 2006; Khairnar et al. 2009; Soares et al. 2012). The strains mainly used for pectinase production studies were *Aspergillus* sp., *Penicillium* sp., and *Thermoascus aurantiacus* (Esawy et al. 2013). This chapter describes the pectinase enzymes and their biochemical characteristics, fungal pectinase production, and the widespread applications of pectinases in processing industries such as wine, fruits, vegetables, textile, tea, and animal feed.

6.2 Biochemical Characteristics of Pectinases

6.2.1 Pectinases

Pectinase or pectinolytic enzyme is a heterogeneous group of enzymes which is used in hydrolyzing the pectic substances. Pectinase is comprised of many groups or types (Sharma et al. 2013; Garg et al. 2016; Kubra et al. 2018; Guo et al. 2019). It is classified into three major groups which are pectinesterases, protopectinase, and depolymerizing enzymes (Kashyap et al. 2001; Kubra et al. 2018). Furthermore, pectinase group can be classified according to their preferred substrate (pectin, pectic acid, or oligo-D-galacturonate), the degradation mechanism (trans-elimination or hydrolysis), and the type of cleavage (random terminal endo or exo) (Favela-Torres et al. 2006; Kubra et al. 2018). The major part of pectinases can be summarized in Table 6.1. Numerous studies were extensively done on the production of pectinolytic enzymes for the effect of different carbon sources and strains (Said et al. 1991; Channe and Sitewale 1995; Arotupin 2007; El-Enshasy et al. 2018; Guo et al. 2019). Among all strains, *Aspergillus niger* is commonly used for the industrial production of pectolytic enzymes (Esawy et al. 2013; El-Enshasy et al. 2018; Sandri and Moura da Silveira 2018; Abdullah et al. 2018). This fungus synthesizes polygalacturonase, polymethylgalacturonase, pectin lyase, and pectinesterase. According to Debing et al. (2006), polygalacturonase, pectinesterase, and pectin lyse are contributing to the breakdown and modification of pectins from a wide variety of plant materials including the citrus pectin.

6.2.1.1 Mode of Action of Pectinases

As explained by Hoondal et al. (2002), an enzyme that hydrolyzes the pectic substances is usually known as pectinases and is comprised of polygalacturonases, pectin esterases, pectin lyases, and pectate lyases depending on their mode of action (Fig. 6.1). In view of this fact, this study focused on *A. niger* that synthesizes pectinolytic enzymes polymethylgalacturonases (PMG) and polygalacturonases (PG); thus, the mode of action of pectinase towards the pectic substance is called as depolymerization (Pedrolli et al. 2009). By referring to Table 6.1, the mechanism of hydrolysis takes part for the PMG and PG since hydrolysis is known as the chemical reaction in which water reacts with a compound to produce other compounds or process that involves the splitting of a bond and the addition of the hydrogen cation and the hydroxide anion from the water. In mechanism, the PMG and PG react towards the pectic substrates and break the glycosidic linkage. The action of the enzyme may be varied depending on its pattern, whether it is random or terminal, and can be defined as endo or exo (Pedrolli et al. 2009).

Table 6.1 Three major types of pectinases

1. Pectinesterases (PE) (EC 3.1.1.11)	Kashyap et al. (2001), Pedrolli et al. (2009), Guo et al. (2019), Oumer (2017), Verma et al. (2018), Kubra et al. (2018), Kavuthodi and Sebastian (2018)
Also known as pectin methyl hydrolase, pectin methoxylase, pectin demethoxylase, and pectolipase	
Catalyzes deesterification of the methoxyl group/residues of pectin and forming pectic acid	
Acts preferentially on a methyl ester group of galacturonate unit next to a nonesterified galacturonate acid	
Pectin + $H_2O \rightarrow$ Pectate + CH_3OH	
2. Protopectinase (EC 4.2.2.2)	
Protopectinase is an enzyme that solubilizes protopectin forming highly polymerized soluble pectin	
Was classified into 2 types:	
(a) Reacts with the polygalacturonic acid region of protopectin, A type	
(b) Reacts with the other polysaccharide chains that may connect the polygalacturonic acid chain and cell wall constituents, B type	
Protopectin (insoluble) + $H_2O \rightarrow$ pectin (soluble)	
3. Depolymerizing enzymes	
Consists of enzyme:	
1. Hydrolyzing glycosidic linkages that include:	
(a) Polymethylgalacturonases (PMG). It catalyzed the hydrolytic cleavage of α-1-4-glycosidic bonds in pectin backbone and maybe endo-PMG (E.C. 3.2.1.15) or exo-PMG (EC 3.2.1.67)	
(b) Polygalacturonases (PG) (EC No 3.2.1.15). It catalyzed the hydrolysis of α-1-4-glycosidic linkage in polygalacturonic acid and may be endo-PG (EC 3.2.1.15) or exo-PG (EC 3.2.1.67)	
2. Cleaving. Cleaving α-1-4-glycosidic linkages by trans-elimination which results in the unsaturated bond between C4 and C5. It may include:	
(a) Polymethylgalacturonate lyases (PMGL) (endo-PMGL or exo-PMGL)	
(b) Polygalacturonate lyses (PGL) (endo-PGL (EC 4.2.2.2) or exo-PGL (EC 4.2.2.9))	

6.2.2 Pectinase Characterization and Purification

Studies of enzyme stability showed the importance of structure and function of enzymes and how to overcome the major restraint for the rapid development of biotechnological processes. Therefore, selection and protein engineering of enzyme structure were carried out to enhance the enzyme stability to a desired levels to be suitable for more industrial applications. As reported by Gummadi and Panda (2003), the physical parameters including pH and temperature are influencing the stability of pectinases. Kashyap et al. (2000) has found that there are two types of pectinases which are acidic and alkaline where the acidic pectinases come from the fungi and alkaline pectinases produced by alkalophilic bacteria, mainly from

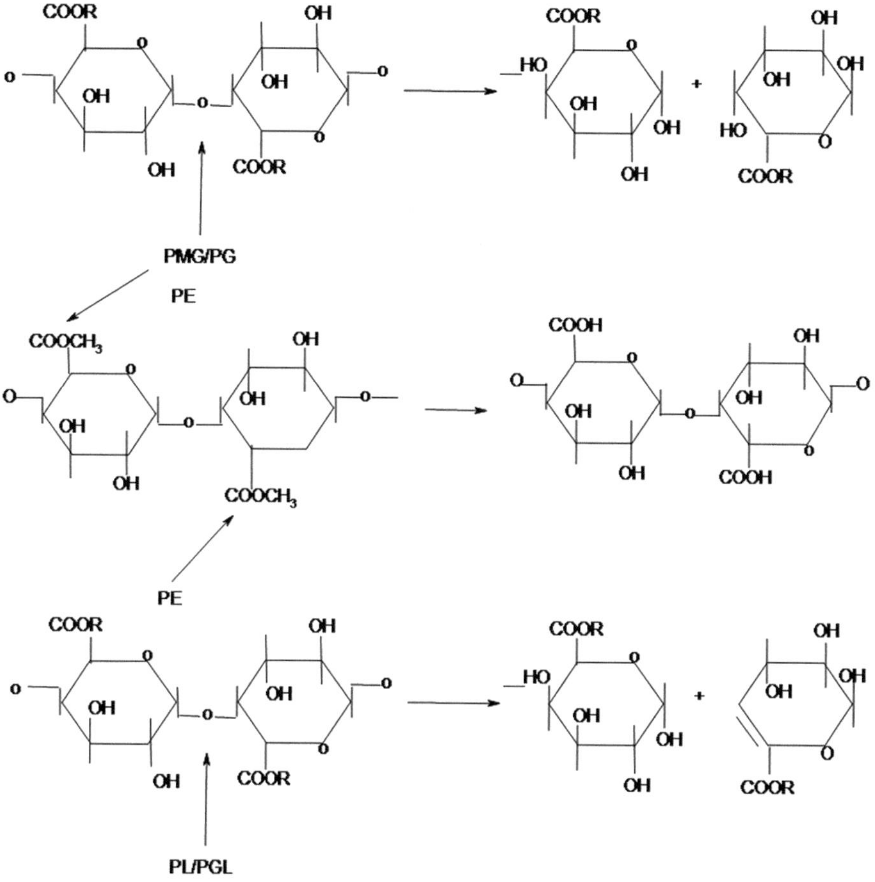

Fig. 6.1 Mode of action of pectinases: (**a**) R = H for PG and CH₃ for PMG; (**b**) PE; and (**c**) R = H for PGL and CH₃ for PL. The arrow indicates the place where the pectinase reacts with the pectic substances. *PMG* polymethylgalacturonases, *PG* polygalacturonases, *PE* pectinesterase, *PL* pectin lyase (Jayani et al. 2005; Pedrolli et al. 2009)

Bacillus species. Joshi et al. (2011) studied on the pH and temperature of the acidic pectinases (pectin methyl esterase, PME) from *Aspergillus niger* and showed that at 30 °C, pectinase activity remained stable and constant up to 6 h at 40 °C and 2 h at 50 °C for all incubation time and decreased for other temperature, whereas at pH 3.5, the maximum activity of pectinase was achieved and the optimal pH was varied from 3.8 to 9.5 depending on the type of enzyme. This is supported by a study from Naidu and Panda (1998) where they found that the optimum pH and temperature for acidic pectinases, PMG, PG, pectin lyase (PL), were 2.23 and 23 ° C, 4.8 and 28 °C, and 3.9 and 29 °C, respectively. In one study conducted by Kashyap et al. (2000), alkaline pectinase produced by *Bacillus* sp. achieved stability of pH and temperature at neutral pH (7.2) with the incubation temperature of 37 °C, and shifting the pH of the medium to acidic (5.0 ± 6.5) or alkaline (8.0 ± 9.0) will decrease the production of pectinases for almost 50%. A study by Khatri et al. (2015) showed that pectinase was active at a temperature between 30 °C and 70 °C and pH (6.2–9.2). The optimum temperature and pH were confirmed at 50 °C and 8.2, respectively. In addition to that, the thermostable and alkaline pectinase was stable up to 70 ° C and about 82% of the enzyme was still active at 100 ° C. Okinji et al. (2019) reported a purified 4.45-fold of pectinase obtained from *Aspergillus fumigatus*, isolated from soil with a recovery of 26.16% with molecular weight of 31.6 kDa. The purified enzyme showed high activity at 60 °C with optimum pH of 5.0 being stable with temperature range between 40 and 50 °C.

Fungal pectinases should remain purified for the characterization of its properties. Pectinases from a wide range of microorganisms have been purified by several researchers (Table 6.2). According to Contreas and Voget (2004), about 470-fold of the PGI was successfully purified with a recovery of 8.6% from a crude sample of *Aspergillus kawachii*. The purification was done using acetone, precipitation, followed by Sepharose Q and Sepharcyl S-100 column chromatography. Different types of microorganisms, pectinase enzyme production, and purification method are presented in Table 6.2.

6.2.3   Chemical Structure and Sources of Pectic Substance

Pectin substances are complex with high molecular mass glycosidic macromolecules comprised of the structure of homogalacturonan (HG), rhamnogalacturonan I (RG-I), rhamnogalacturonan II (RG-II), and xylogalacturonan (XGA). Homogalactron (HG) is the simplest form of pectin because it is a linear polymer composed of 1,4-linked α-D-galacturonic acid (GalA) residue (Martens-uzunova and Schaap 2009; Pedrolli et al. 2009). However, pectin have been also found in other polymer form composed of HG and RGI part with side chain composed of neutral sugars (Voragen et al. 1995; Pedrolli et al. 2009). Although the structure of rhamnogalacturonan II (RG-II) is less abundant, it is the most complex form of pectin structure compared to rhamnogalacturonan I (RG-I). According to Zakharova et al. (2018), the pectin structure is built from a backbone of rhamnogalacturonan I (RG-I) as this structure is the most common form where α-L-rhamnose of the hydroxyl groups of C2 and C1 is inside the polygalacturonic acid chain and linked to the C1 and C4 of the different ends of polygalacturonic acid chain. From the linked C4 of the α-L-rhamnose, the side chains can be varied which are galactran, arabinan, or arabinogalactans I and II. Furthermore, for some elongation or extension of the side chain, the monomer of L-rhamnose will form and are called hairy regions, whereas for some less elongation or extension of the side chain with L-rhamnose, they are called smooth regions. However, the other references from Richard and Hilditch (2009) stated that in pectin, the monomer of D-galacturonic

Table 6.2 Different types of biofactories for pectinase production commonly used for industrial production microorganisms

Strain	Purification method	Recovery	Purification stage
Aspergillus fumigatus (Okinji et al. 2019)	CM-Sephadex C-50 and Sephacryl S-200	26.1%	Enzyme pectinase with 4.45-fold
Aspergillus kawachii Contreas and Voget (2004)	Acetone precipitation	8.6%	Enzyme PGI with 470-fold
	Sepharose Q		
	Sephacryl S-100 column chromatography		
Aspergillus niger MTCC 478	Acetone precipitation	60–90%	Enzyme PG with 15.28-fold
Anand et al. (2017)	CM cellulose column chromatography		
Aspergillus flavus Celestino et al. (2006)	Ammonium sulfate fraction	10.3%	Enzyme PL with 58-fold
	DEAE-cellulose ion exchange		
	Sephadex G-100 gel filtration		
Aspergillus japonicus Semenova et al. (2003)	Hydrophobic	–	Enzyme PGI, PG11, PE1, PE11, PL with range 2–20-fold
	Ion exchange column chromatography		
Aspergillus niger ANO7	Anion exchange chromatography on DEAE-cellulose	52.6%	Enzyme PG with 24.8-fold
Patidar et al. (2017)	Gel filtration chromatography using Sephadex G-200		
Aspergillus flavus Doughari and Onyebarachi (2019)	Ethanol precipitation	3%	Enzyme PG with 9.93-fold
	Sephadex G-75		
Aspergillus awamori L1 Ngo et al. (2008)	Ethanol precipitation	68.6%	Enzyme PG with 30.4-fold
	Sephadex G-75 gel filtration		
Thermoascus aurantiacus	Sephadex G-75 gel filtration	24.6%	Enzyme PG with 21-fold
Martins et al. (2007)	SP-Sepharose ion exchange chromatography		
Acrophialophora nainiana	Sephacryl S-100 gel filtration	60.6%	Enzyme PG and PL with 9.37-fold
Celestino et al. (2006)	DEAE-Sepharose ion exchange		
	Sephadex FG-50		
Streptomyces lydicus Yadav et al. (2008)	Ultrafiltration	54.6%	Enzyme PG with 57.1-fold
	CM-cellulose		
	Sephadex G-100 column chromatography		

Note: *PG* polygalacturonases (EC 3.2.1.15), *PE* pectinesterase (EC 3.1.1.11), *PL* pectin lyase (EC4.2.2.10)

acid is present in the form of α-pyranose hydroxyl group of C1 and C4 and forms a long chain. Therefore, the pectin backbone is composed of major constituent of polygalacturonic acid chain or monomer of D-galacturonic acid (GalA) and also represents a significant carbon source for bacteria or fungi living on the decaying plant material (Martens-Uzunova and Schaap 2008). D-Galacturonic acid (GalA) is an aldose sugar with an acid group or called as a uronic acid (Richard and Hilditch 2009). About 80% of the carboxyl group of D-galacturonic acid will change into methanol in nature and the ratio of methylated D-galacturonic acid will decrease during the extraction of pectin (Richard and Hilditch 2009).

In nature, the pectic substances are omnipresent in the plant kingdom and will form the large components of the middle lamella, and the material that is found between the primary cell walls side by side of young plant cells is comprised of a thin layer of extracellular adhesive (Hoondal et al. 2002). Study conducted by Jayani et al. (2005) showed that the highest pectic substance was found in the orange pulp which was 12.4–28.0%, whereas the lowest pectic substances were found in banana and apple at 0.7–1.2% and 0.5–1.6% (Table 6.3). About 0.4–0.5% of the weight of fresh material consists of pectic substances (Kashyap et al. 2001). Another study from Sharma et al. (2013) stated that pectin is taking one-third of the dry weight of plant tissue. Furthermore, the pectic substances are acidic, very high molecular weight, and negatively charged. Since it is a complex polysaccharide group, the pectic substance does not have an exact molecular weight and can be estimated in about 25–360 kDa (Kashyap et al. 2001). According to Kashyap et al. (2001), in practical application of pectin extraction, when the tissue is ground, soluble pectin is found at the liquid phase and caused the rising of viscosity and the pulp particle. The other pectin molecules will still link to cellulose fibril or side chain of hemicelluloses and will assist in the water retention.

Table 6.3 Pectic substance content in different fruits and vegetables (Jayani et al. 2005; Pasha et al. 2013; Tapre and Jain 2014; Javed et al. 2018)

Fruits/vegetables	Pectic substance (%)	Tissues
Apple	0.5–1.6	Fresh
Orange pulp	12.4–28.0	Dry matter
Orange	29	Fruit
Peach	0.3–1.6	Fruit
Sugar beet pulp	30	Fruit
Banana	0.7–1.2	Fresh
Tomatoes	2.4–4.6	Dry matter
Potatoes	1.8–3.3	Dry matter
Carrots	6.9–18.6	Dry matter

6.3 Pectinase Production by Microbial Cell

6.3.1 Fungal Bioprocessing for Pectinase Production

Over the centuries, biotechnology is emerging with respect to the processes that involved living organisms. A study by Wainwright (1992) revealed the potential and capability of true filamentous fungi with different species in the processes such as production of antibiotics, organic acid, enzyme, traditional food, and others. The organism involved and product produced are shown in Table 6.4. Filamentous fungi are of high interest because of their high production of enzymes and also for other metabolites. In addition to that, the genus of *Aspergillus* has been successfully used

Table 6.4 Industrial significance of products produced by filamentous fungi (Wainwright 1992)

Organism	Product
Antibiotics	
Aspergillus fumigates	Fumagillin
Cephalosporium acremonium	Cephalosporin C
Fusidium coccineum	Fusidic acid
Penicillium sp.	Griseofulvin
Penicillium chrysogenum	Penicillin
Helminthosporium siccans	Siccanin
Organic acid	
Aspergillus niger	Citric acid, gluconic acid
Aspergillus terreus	Itaconic acid
Enzyme	
Aspergillus niger	Lipase, pectinase, glucose oxidase, lactase, pentosanase, proteases, β-glucanase, cellulase, glucoamylase
Aspergillus oryzae	α-Amylase
Aspergillus awamori	Glucoamylase
Trichoderma sp.	Dextranase, cellulose
Penicillium sp.	Dextranase
Mucor sp.	Rennin
Traditional foods	
Aspergillus oryzae	Miso
Mucor, Rhizopus	Ragi
Mixed culture	Shoyu (soy sauce)
Others	
Gibberella fujikuroi	Gibberellins
Gibberella zeae	Zearalenone
Claviceps sp.	Ergot alkaloids
Aureobasidium pullulans	Pullulan
Sclerotium sp.	Scleroglucan
Rhizopus sp., *Fusarium* sp.	Steroid bioconversion

as a production host (Wang et al. 2005). Punt et al. (2002) reported that *A. niger* is commercially used in the food industry and classified as a group of GRAS (generally recognized as safe) in accordance with the Food and Drug Administration (FDA). As shown in Table 6.4, *A. niger* can produce multiple enzymes such as pectinase, lipase, glucose oxidase, lactase, pentosanase, proteases, β-glucanase, cellulase, glucoamylase, and α-amylase.

Different approaches were used in order to optimize the application of these filamentous fungi including from the molecular part and also biotransformation. In a previous research from Miyazaki et al. (2011), a gene for processing the α-glucosidase I from the *Aspergillus brasiliensis* ATCC 9642 for heterologous expression of fusion enzyme hydrolyzed pyridylaminated (PA) oligosaccharides Glc3Man9GlcNAc2-PA and Glc3Man4-PA. This fusion enzyme is an oligosaccharide precursor of eukaryotic N-linked glycoproteins and processing α-glucosidase I. Besides, Abas et al. (2010) studied the effect on the biotransformation of (*R*)-1-(4-bromo-phenyl)-ethanol by using the *Aspergillus niger* as biocatalyst. The usage of *Aspergillus niger* as biocatalyst in the studies is due to the simple function of asymmetric reduction of enzyme 1-(4-bromo-phenyl)-ethanone to (*R*)-1-(4-bromophenyl)-ethanol through the optimized condition that influenced the bioconversion.

With regard to the capability of fungal applications in many industrial processes, fungal is very much desired for the production of the enzyme pectinases. There are many studies that have been reported for the efficiency of fungal used as biofactories for pectinase production. Ahmed et al. (2016) reported the production of pectinase by *Aspergillus niger* using citrus waste peel. The maximum enzyme yield of 117.1 mM/mL/min was obtained in an orange waste peel medium and optimally active at pH ¼ 7 and 55 °C. Another study by El-Enshasy et al. (2018) reported the bioprocess optimization for pectinase production using *Aspergillus niger*. They found that the enzyme production was able to be produced up to 450 U/mL after 126 h cultivation time in submerged cultivation using fed-batch strategy in a bioreactor. A recent study conducted by Mehmood et al. (2019) showed maximum pectinase production of 480.45 U/mL within a period of 1-day cultivation under solid-state fermentation. The production of pectinase is high and within a short period of time; however, it required further study in terms of large-scale production using solid stage cultivation strategy.

6.3.2 D-Galacturonic Acid Degradation Pathway

It is well known that D-galacturonic acid is the main constituent of pectin. It also is a natural and plentiful component of polysaccharides and can also serve as a pivotal carbon source for microbes (Zhang et al. 2011). There are two types of backbones present in pectin which are homogalacturonan (smooth region) and rhamnogalacturonan I (hairy region). Homogalacturonan consists of α-1,4-linked D-galacturonic acid, whereas rhamnogalacturonan I comprises alternating α-1,4 linked

D-galacturonic acid and α-1,2-linked rhamnose residue (Richard and Hilditch 2009). In the "hairy region," xylogalacturonan comprising of D-xylose-substituted galacturonan backbone and rhamnogalacturonan can be recognized (de Vries and Visser 2001).

In fungus, D-galacturonic acid is catabolized through a pathway with four enzymes and two reduction steps, whereas five enzymes are involved in catabolic pathway for bacteria (Khosravi et al. 2015). The Initial steps of degradation in eukaryotes invovle NADPH-dependent D-galacuronic reductase which is coded as gaaA in *A. niger* or gar1 in *T. reesei* which govern the conversion of D-galacturonate to L-galacturonic acid by utilizing NADPH or NADH as electron doner (Kuorelahti et al. 2006). In this study, D-galacturonic reductase can use NADPH and NADH but it showed a higher affinity to NADPH (Richard and Hilditch 2009). L-Galactonic acid serves as intermediate in the eukaryotic route for D-galacturonic acid metabolism (Kuivanen et al. 2012). L-Galactonate dehydratase (LGD1) coded as gaaB in *A. niger* further converts L-galactonic acid to 2-keto-3-deoxy-L-galactonic acid (Biz et al. 2016). Water molecules from L-galactonic acid were split by dehydratase to form this molecule (Kuorelahti et al. 2006). The third step in the pathway is catalyzed by 2-keto-3-deoxy-L-galactonate aldolase coded as GaaC in *A. niger* which cut 2-keto-3-deoxy-L-galactonate between carbon 3 and 4 into pyruvate and L-glyceraldehyde (Biz et al. 2016). Pyruvate can be utilized by different pathways like the citric acid cycle, while L-glyceraldehyde cannot be utilized in any known pathway. However, a specific NADPH-dependent glyceraldehyde reductase coded as gaaD in *A. niger* has been reported in conversion of L-glyceraldehyde to glycerol using NADPH as electron donor (Khosravi et al. 2015; Biz et al. 2016). In the eukaryotic route, the second and fourth enzymes are activating only for one direction, whereas the first and third enzymes are reversible (Richard and Hilditch 2009).

Considering pectin as a crucial carbon source for bacteria and also fungi living on decaying plant material, the metabolic pathway was concentrated on the main backbone of pectin which was D-galacturonic acid (GalA) (Martens-Uzunova and Schaap 2008). The metabolic pathway of filamentous fungi was different from the bacteria, and a non-phosphorolytic pathway was proposed, leading to glyceraldehydes and pyruvate as end product of the pathway (Martens-Uzunova and Schaap 2008) (Fig. 6.2). A study from Sealy-Lewis and Fairhurst (1992) showed in filamentous fungus that an NADPH-dependent D-glyceraldehyde (D-GAD) reductase was induced on D-galacturonate and an $NADP^+$-dependent glycerol dehydrogenase was also then discovered to induce on D-galactonate and helped in reducing glycerol.

6.3.3 Heterologous Pectinase Production

In order to meet the increasing demand for the production of pectinase enzyme, there are several approaches applied in its production. There have been a number of longitudinal studies in increasing the production through optimization methods as

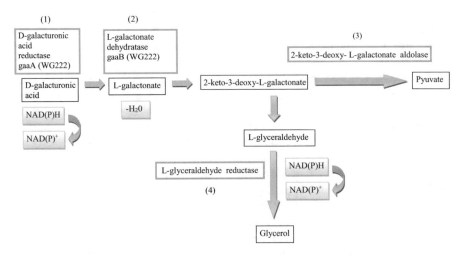

Fig. 6.2 Proposed metabolic pathway of galacturonic acid degradation in filamentous fungi. (*1*) D-galacturonic acid reductase (GAA), (*2*) L-galactonate dehydratase (GAAB), (*3*) 2-keto-3-deoxy-L-galactonate aldolase (GAAC), and (*4*) L-glyceraldehyde reductase (GAAD). Indicated also the gaa loci and mutant strains of filamentous fungus of *Aspergillus nidulans*. (Modified from Martens-uzunova and Schaap 2008)

well as applying a semi-industrial scale that has been reported by Farid et al. (2000) and El-Enshasy et al. (2006, 2018).

Current research on expression systems is focused on two general categories which are prokaryotic and eukaryotic systems. The advantages of using prokaryotic system were they are easier to handle and applicable for most purposes, but unfortunately, the prokaryotic system has limitations for the production of eukaryotic protein especially in post-translational modification like proper folding, glycosylation, phosphorylation, and formation of bridges (Rai and Padh 2001). Hence, choosing the best expression system in protein expression should be based on some options like yield to glycosylation, proper folding, and the economics of scaling up. Study by Porro et al. (2011) showed the comparison of different host cell features in terms of quantity and quality of the production (Table 6.5). In order to meet the required production of industrial enzymes, the high yield of production must be reached.

The filamentous fungi were chosen as a production host instead of yeast or *Escherichia coli* for the production of protein due to many factors such as their ability to grow on low-cost medium and their high secretion capacity to the culture medium. Research by Nevalainen et al. (2005) showed that the natural ability of filamentous fungi secretes a large amount of protein in the growth medium compared to the other microorganisms.

According to Wang et al. (2005), there are many bioprocessing parameters which include cultivation medium and effect on condition (pH, agitation, and shear stress) that should be taken into account in order to improve the yield in terms of growth and production. In a study conducted by Solis-Pereyra et al. (1993), it is shown that

Table 6.5 Comparison between different host cell cells in terms of cell growth rate, product concentration, and location of product

	Location of product inside the cell	Biomass cell dry weight (gL⁻¹)	Typical spec. growth rate (gL⁻¹)	Product concentration (gL⁻¹)	References
B. subtilis	Secreted	20–35	0.05–0.8	≤3	Westers et al. (2004)
E. coli	Cytoplasm	50–70	0.08–0.12	10–15	Tripathi et al. (2009)
E. coli	Secreted	20–60	<0.1	0.2–2	Georgiou and Segatori (2005)
S. cerevisiae	Cytoplasm	<90	0.01–0.2	0.5–1.5	(Knoll et al. 2007)
	Secreted	60–100	0.01–0.15	≤1	
P. pastoris	Cytoplasm	80–150	0.02–0.04	10–20	Jahic et al. (2002)
	Secreted	80–150	0.02–0.18	<5	
A. awamori	Secreted	–	–	4.6	Schmidt (2004)

the synthesis of pectinase by microorganism was highly affected by the growth medium components which include sources from carbon, nitrogen, and other elements like mineral. Based on the idea of Papagianni (2004), there are at least 13 elements for essential growth including oxygen, manganese, iron, zinc, copper, and molybdenum. Lacking one of those sources may retard the growth condition of microorganisms and will decrease the production of the enzyme. Hence, the sufficiency of the elements in the cultivation medium was crucially important. From the previous study by Suhaimi et al. (2016), they found that every component such as carbon and nitrogen sources gives positive influences towards the cultivation medium and with correct type and concentration of carbon and nitrogen sources enhanced the production of pectinase enzyme.

Carbon compounds can range from simple small molecules like sugars, organic acids, alcohol, polysaccharides, and lipids. Usually, the industrial enzyme such as pectinases will use the complex media with a solid substrate for cultivation (Papagianni 2004). It is because the complex media with the solid substrate highly influenced the morphology and growth kinetics of the microorganism. Fungi are organisms that cannot fix nitrogen (Kavanagh 2005). Thus, a compound containing nitrogen needs to be supplied either in organics such as amino acid or inorganic form such as ammonium sulfate. The microbial cell utilized the nitrogenous compounds for the synthesis of nucleotides, amino acid, protein, enzyme, and other metabolite (Arijit et al. 2013).

As explained by Said et al. (1991), there are various carbon compounds examined as an inducer for pectinase production on *Penicillium frequentans* which include pectin, polygalacturonic acid, and citrus pulp pellets, and it is proved that pectin is a good substrate/inducer for enzyme production (Table 6.6). These results are in a good agreement with other studies (Aguilar and Huitrón 1990; Arotupin 2007) which have shown the importance of the presence of pectin in culture medium for better enzyme production.

Table 6.6 Effect of different carbon sources on pectinase production and growth of *Penicillium frequentans*

Carbon source (0.5%)	Reducing group-releasing activity/Pectinase (μmol^{-1} min^{-1} mL^{-1})	Cell dry weight (mg)
Galacturonic acid	0.10	12.2
Polygalacturonic acid	0.56	15.8
Pectin	1.20	17.4
Citrus pulp pellets	0.41	–

Source: (Said et al. 1991)

Table 6.7 Pectinase production genes

Gene	Corresponding enzyme	Probe fragment	CCC TGA box	References
pelA	Pectin lyase A	1.6 kb *Cla*I	+	de Vries et al. (2002)
pelB	Pectin lyase B	1.1 kb *Sma*I	+	de Vries et al. (2002)
pelC	Pectin lyase C	3 kb *Nsi*I	ND	de Vries et al. (2002)
pelD	Pectin lyase D	0.7 kb *Xho*I	+	Harmsen et al. (1990)
pelF	Pectin lyase F	0.9 kb *Sst*I	ND	Pařenicova et al. (2000)
pgaA	Polygalacturonase A	0.33 kb *Xho*I/*Xba*I	+	Bussink et al. (1992)
pgaB	Polygalacturonase B	0.20 kb *Nsi*I/*Kpn*I	+	Pařenicová et al. (1998)
pgaC	Polygalacturonase C	0.28 kb *Nsi*I/*Sma*I	+	Pařenicová et al. (1998)
pgaD	Polygalacturonase D	0.13 kb *Cla*I/*Sst*I	ND	de Vries et al. (2002)
pgaE	Polygalacturonase E	0.28 kb *Nsi*I/*Xho*I	+	
*pga*I	Polygalacturonase I	0.28 kb *Nsi*I/*Xba*I	+	
*pga*II	Polygalacturonase II	0.38 kb *Nsi*I/*Nco*I	+	
pgaX	Exopolygalacturonase	1.7 kb *Pst*I	ND	
plyA	Pectate lyase A	0.9 kb *Pst*I/*Xho*I	ND	
pmeA	Pectin methyl esterase	0.86 kb *Nsi*I/*Xho*I	ND	

6.3.4 Gene Regulator for Fungal Pectinase

Polygalacturonase (PG), pectin lyase (PL), and pectin methyl esterase (PME) encoding a gene expression through glucose repression were induced from a source containing pectin, polygalacturonic acid, and D-galacturonic acid (de Vries and Wackernagel 2002). Hence, a similar study also recognized that the mostly involved in repression of pectinolytic gene expression was carbon catabolite repressor protein (CreA). In addition, the promoter used in the pectinolytic gene of *Aspergillus niger* was in a region (5′PTYATGGTGGAA3P′) in the presence of D-galacturonic acid as inducer, whereas its in the (CCTGA) region in the presence of L-arabinose.

From the study of de Vries et al. (2002), 26 pectinolytic genes of *A. niger* in a wild type strain were studied based on a detailed analysis, which included hybridization probe and in silico analysis of CCTGA box, of their expression on pectin and pectin-derived carbon sources. However, in this review, only pectinases with corresponding gene will be shown (Table 6.7).

6.3.5 Recombinant Pectinase Production by Fungi

The recombinant DNA technology has some of the efficient tools like gene cloning and mutation which attracted significant interest in carrying out research studies in the modern era of biotechnology for the production of useful enzymes. By means of these incredible tools, scientists not only can study the specific gene of interest but also can achieve targeted or overexpression of a specific protein for some particular interest (Alimardani et al. 2011). Pectinase has been studied for a very long time aiming to retrieve a large-scale expression due to its potential for various industrial applications. Recombinant technology has provided the different types of techniques to produce this particular enzyme of interest on a commercial scale in cost-efficient manner. Various pectinolytic genes have been cloned and expressed in the vast range of host organisms in order to obtain their homologous as well as heterologous expression with very minimum energy, cost, and time (Barbosa et al. 2015).

According to the study conducted by Gonçalves et al. (2012), the genetically modified version of *P. griseoroseum* was used to gain a high production of pectinases. This recombinant was constructed and coupled with additional copies of the pectinase genes which resulted in 13 folds and 424 folds of hyperproduction of pectin lyase and PGase. In recent times, the expression of pectinase from *Aspergillus* in *P. pastoris* results in high activity in culture broth using high-cell-density batch fermentation, and the yield was 8 times greater than the corresponding shake flask cultivation. Some studies have been conducted for the expression of acid-tolerant endopolygalacturonase gene from *P. oxalicum* to the host of *P. pastoris*. By doing this, the yield of pectinase enzyme was maximized up to 1828.7 U/mL. When constitutive promoter such as the promoter of gpd gene is used to substitute the indigenous promoter, the production level of pectinase significantly increases.

Besides that, the UV-induced spontaneous catabolite repression-resistant mutant of *P. griseoroseum* led to the elevated production level up to 7–8 folds of pectinase titer when compared to its native strain (Li et al. 2017). The expression of the pelB gene in *E. coli* is noteworthy as further optimization on this recombinant strain in 7 liters bioreactor can enhance the pectinase titer up to 1816.2 U/mL. Most of the industrial enzymes are being produced at a large scale from the source of the fungus. Most of the fungal strains secrete enzymes similar to other similar proteins (Asgher et al. 2016). According to the study conducted by Hadj-Taieb et al. (2006), when the strain *P. occitanis* genetically modified with the single round of nitrous acid mutagenesis, the mutant strain secretes 50 times more endo- and exo-pectinase enzyme, unlike the wild-type strain which relatively produces lesser amount of pectinase enzyme. A cloned pectate lyase gene from *Klebsiella* sp. was expressed in *E. coli* BL21 strain which later was grown in LB medium. The culture supernatant and the cell lysate showed around 4.5 U/mL and 1.2 U/mL of pectinase enzyme level (Yuan et al. 2011). Yang et al. (2011) also reported that the clone of endopolygalacturonase gene (pga1) from the acidophilic fungus of *Bispora* sp. was expressed in *P. pastoris* GS115 by using the pPIC9 as the expression vector. They transformed *P. pastoris* then only cultivated in yeast extract peptone dextrose medium at

30 °C. Pectinolytic enzymes were extracted successfully with methanol induction at the gain of 50 U/mL.

6.3.6 Parameters Affecting Enzyme Production

Production of pectinase is in high demand in industry. Various agro-industrial wastes are rich in polysaccharides, especially pectin, which are being produced in large amounts. Some of these wastes can be used for the production of pectinase enzymes. Pectinases are produced by several microorganisms, particularly fungal strains which are known as good producers. *Aspergillus* and *Penicillium* strains are widely exploited for pectinase production at the industrial scale, by submerged fermentation as well as solid-state fermentation. Besides the selection of suitable microbial strains (wild type, recombinant, mutagenized), various other parameters also affect pectinase production.

6.3.6.1 Fermentation pH Value

Fermentation pH value greatly affects the stability of the pectinase enzyme production. The amounts of pectinesterase I (PE I), pectinesterase I1 (PE 11), polygalacturonase (PG), and pectin lyase (Pl) produced from microbial fermentation are greatly affected by the pH value of the medium. Most of the PGase enzymes are reported with acidic optimum pH in the range of 3.3–3.7 (Ramadan 2019). Fungi and yeast produce PGase with acidic pH. However, the production of extracellular PGases from *Fusarium oxysporum* and intracellular PGases from *Bacillus licheniformis* is effective under alkaline conditions (Kaur et al. 2010). The optimum pH for growth and pectinase production for most of the bacteria is 7–10. The production of alkaline pectinase from *Bacillus* sp. Y1 is obtained in a statistically optimized medium with pH of 8.2 (Guo et al. 2019). Xylano-pectinolytic enzymes find a wide application in biobleaching industry, and the optimal efficiency is at pH 8.5 (Singh et al. 2019).

6.3.6.2 Temperature

The production of pectinases by microbial fermentation also depends on thermal stability. In most of the fruit juice and wine production industries, pectinases from *Mucor rouxii* are efficient at 20 °C. A slight increase to 30 °C is sensitive and could affect the quality of beverages produced (Ramadan 2019). The optimum temperature for pectinase production by *Bacillus subtilis* was found to be 37 °C. Pectinase enzyme can give maximum juice yield, i.e., 92.4%, at an incubation temperature of 37 °C for 360 min with 5 mg/100 g of enzyme concentration, whereas the combination of two enzymes, i.e., pectin methyl esterase (PME) and polygalacturonase (PG) at 120 min of incubation time, 50 °C of incubation temperature, and 0.05 mg/100 gm of enzymatic concentration can give the maximum yield of 96.8% for plum fruits (Sharma et al. 2017). PGase from fungi has optimum activity at 50 °C, while from yeast the temperature varies from 40 °C to 60 °C (Sandri et al. 2011).

6.3.6.3 Effect of Metal Ion

The production of microbial enzymes is also affected by metal ions. Studies reported that the presence of Cu^{2+} and Hg^+ greatly reduces the activity of endo PGase. Metal ions such as Hg^{2+}, Zn^{2+}, and Mg^{2+} inhibit enzyme production due to inhibition by thiol group blocking agents as there is possible involvement of the thiol group in the enzyme's active site (Kaur et al. 2010; Guo et al. 2019). The presence of Mn^{2+} could increase the PGase activity. However, Li^+, Fe^{2+}, and Rb^{2+} have no effect on the activity. At high concentration of metal ions, the enzyme production is low due to blockage of secretion of protein into external medium. Bacterial pectate lyases need Ca^{2+} for growth, whereas fungal pectinase does not need Ca^{2+} (Sandri et al. 2011; Sharma et al. 2017).

6.3.7 Pectin as an Inducer

Microbial systems enhance pectinolytic production using pectin as an inducer. The main chain of pectin is methylated 1,4-D-galacturonan molecule. Demethylated pectin is known as pectic acid or polygalacturonic acid. Pectic substances are classified into four main types: pectic acids, pectinic acid, protopectin, and pectin. Pectic acid is a group designation applied to pectic substances mostly composed of colloidal polygalacturonic acids and essentially free from methoxy groups. The salts of pectic acid are either normal or acid pectates. The pectinic acids contain up to 75% methylated galacturonate units. Under suitable conditions, pectinic acids are capable of forming gels with sugars and acid or, if suitably low in methoxyl content, with certain metallic ions. The salts of pectinic acid are either normal or acid pectinates. Protopectin is the water-insoluble parent pectic substance, located primarily in the middle lamella that serves as the glue to hold cells together in the cell walls. It yields pectin or pectinic acids upon restricted hydrolysis.

Pectic substances are commonly amorphous, with a degree of polymerization of about 200–400. Substituents can be found at the C-2 or C-3 positions of the main chain. Substituents can be either non-sugar (acetyl) or sugar (D-galactose, D-xylose, L-arabinose, and L-rhamnose). The degree and type of branching varies depending on the source of the pectic substance. The synthesis of pectic substances occurs in the Golgi apparatus from UDP-D-galacturonic acid during the early stages of growth in young enlarging cell walls. Lignified tissues have very lower pectin content compared with young and actively growing tissues. Higher plants usually have very low pectic content, especially less than 1%. Vegetables and fruits are abundant resources for pectin. Deng et al. (2019) reported that ripeness of fruits had significant influence on the drying kinetics, which is related to the modification of physicochemical and pectic properties. The water-soluble pectin content increased as ripening progressed, but the chelate- and sodium carbonate-soluble pectin contents gradually declined (Deng et al. 2019).

Now, the most common raw materials for the production of pectin are the residues from the manufacture of fruit juices, including apple pomace and citrus fruits (Alkorta et al. 1998; Blanco et al. 1999). The extraction of pectin is carried out by acid hydrolysis at a pH range of 2–3 for 5 h at high temperature (70–100 °C). The solid-to-liquid ratio is normally about 1:18. The pectin extract is separated from the pomace using a hydraulic press or by centrifugation. The extract is filtered and finally concentrated on a standard-setting strength. For powdered pectin preparation, the concentrated liquor is treated with organic solvents or certain metallic salts to precipitate the polymers (Sakai et al. 1993).

Agro-industrial residues which are primarily composed of complex polysaccharides are also utilized to strengthen microbial growth for the production of industrially important pectinases. During the raw agriculture material processing for food, a bulk quantity of agro-wastes is generated. Among these agro-residues, dried citrus peel and orange bagasse contain large amounts of soluble carbohydrates, particularly fructose, glucose, sucrose, and pectin. Citrus peel is known as the major source of pectinases in several industries instead of apple pomace. A significant quantity of pectin in citrus peel urges us to use them as a substrate for microorganism to produce pectinolytic enzymes (Ahmed et al. 2016).

6.4 Pectinase Applications

6.4.1 Wine Industry

Wine is a word that evokes emotion in millions across the world. Red and white wine are fermented drinks highly appreciated worldwide as they present a very complex composition. They contain a number of specific compounds which are polyphenols, anthocyanins, polysaccharides, and volatile compounds. In order to achieve the best wine properties, efficient treatments must be very complex in all wine production steps such as maceration and clarification. Within this context, utilization of the enzymes in different steps of winemaking is crucial (Chandrasekaran et al. 2015). Pectic enzymes are utilized in winemaking from different fruit materials. These enzymes are used at various stages of the winemaking process, especially the crushing of fruits. Enzyme treatments during the crushing of fruits will increase the yield of the juices and reduce the pressing time. In addition, enzyme treatment helps to improve the color yield due to the extraction of pigments. Furthermore, the addition of enzyme before or during fermentation helps to settle out undesirable microbes with suspended particles. As a result, a better quality of wine is being produced (Nighojkar et al. 2019). Grapefruits are commonly used to produce wine. Grapefruits contain various nutritious compounds such as anthocyanins, tannins, stilbenes, and aromatic terpenes which are mainly localized in the vacuole of grape skin cells that confer the health benefits to the host. The efficiency and control of maceration process of the grapefruit cell wall are playing a pivotal role in releasing these desirable compounds (Gao et al. 2016). Hence, the overall color intensity and stability of wine can be enhanced by triggering anthocyanins to bind with the tannin as well as its structure and phenolic extraction of red wines (Mojsov et al. 2015). Maceration during fermentation is very crucial as the extraction of these beneficial metabolites and macromolecules is being used in alcoholic fermentation during conversion to wine (Gao et al. 2016).

In wine production, the use of pectinolytic yeasts or their enzymes could be highly beneficial in the development of wine fermentation process of grape juice and overall wine production. Traditionally, the three main enzymes used in wine-making are pectinases, β-glucanases, and hemicellulases. These enzymes served as the polysaccharide degradation in cell wall to heighten the maceration process of grape skin. Hence, extraction of natural color from grape skin can be greatly enhanced. In addition, clarification and filtration processes can be greatly enhanced to achieve a good grade of wine quality (Toushik et al. 2017). This is usually carried by applying pectinase which help in releasing the entrapped color, flavor, and phenolic compounds. The pectinase is used in wine to maintain low level of methanol (Fernández-González et al. 2004). In addition, it can aid in starch hydrolysis (alpha-amylases), thus improving the filtration problems caused by beta-glucans present in malt, hydrolyzing proteins, and controlling haze during maturation, filtration, and storage (Patel et al. 2016). Furthermore, the enzyme helps in pre-fermentative cold soak process that stabilizes the color and flavor of wines (Merín and Morata de Ambrosini 2015). Additionally, enzyme treatment helps to settle particle more rapidly. The action of pectinase on negatively charged pectin molecules exposes positively charged grape solid, leading to increment of flocculation (Mojsov et al. 2015).

6.4.2 Fruit and Vegetable Processing

Enzymes are pivotal throughout the life cycles of fruits and vegetables. However, the enzyme is still remaining active even after harvesting, and this contributes to the undesirable effects on texture, smell, taste, and nutritional value (Bayindirli 2010). The final quality of fruits and vegetable production can be manipulated using enzymes by disabling the unwanted enzymes that bring undesirable effects (Chandrasekaran et al. 2015). The enzymes are necessary for reducing the viscosity of concentrates, removing cells from plant material that enhances the yield of juice and solids, and modifying and solubilizing pectic structures to affect sedimentation and clarification of juices (Nighojkar et al. 2019). Over the past decade, fruits and vegetables received gigantic attention as the market and health awareness of the people rapidly grew (Hyson 2015). In general, fruit industries utilized various enzymes to improve the quality of the products and reduce the costing of production (Chandrasekaran et al. 2015).

Pectinase enzyme is applied in the fruit juice industry to decrease the viscosity, increase the yield and clarification of juice by liquefaction of pulps, and remove the peels of the fruit. In this process, cell walls will be degraded by pectinases and increase the yield of the juice. In addition, it helps in the maceration of vegetables to produce pastes and purees (Garg et al. 2016). Clarification of fruit juice is one of the crucial applications of acidic pectinase. Generally, the colloids are polysaccharides contained inside the fruit juice that contribute to the fouling problem during the filtration process (Rai et al. 2004). Therefore, pectinases are utilized for pretreatment of juices to reduce the number of polysaccharides like pectin and starch. Hence, the viscosity of the juice can be reduced and speed up the filtration process and enhance the clarity of juice (Oumer 2017). This is due to the crushing of pectin-rich fruit juice which stays linked to the fruit pulp in gelatinous structure and hinders the juice clarification process by pressing results in high viscosity juice. Thus, the addition of pectinase will degrade the gel structures and enhance the juice concentration capacity (Pedrolli et al. 2009). The enzyme treatment can reduce viscosity up to 62%. Compared to undepectinized juice, after ultrafiltering of depectinized apple juice, the permeate flux is much higher. The accretion of permeation rate indicates the viscosity and total amount of pectin content is being reduced (Alvarez et al. 1998). The addition of pectinase enhances the phenolic content being released from the fruit skin. These phenolic contents contribute to antioxidant effects, conferring the health benefit to the host (Garg et al. 2016).

According to the studies, different fruit juice needs a different type of enzyme treatment. For example, apples, berries, and grapes need an intense treatment with a combination of pectinase and cellulases for complete clearance of haze, turbidity, and viscosity. Fruit juices like orange and pineapple prefer mild pectinase treatment for the extraction of juice as the main target is to have a high yield of juice but no intense clarification as these juices are made for the consumer. Enzyme treatment also improves visibility by preventing the darkening of juice as the unwanted haze due to pectin fibers is removed. Hence, the color of the fruit juice can be preserved (Singh 2018).

6.4.3 Coffee Processing

The fermentation of coffee using pectinolytic microorganism removes the mucilage coat from the coffee beans (Oumer 2017). In general, mucilage from coffee beans by natural fermentation is removed using water. However, the mucilage breakdown is always not complete even after long hours of fermentation. If the coffee beans are fermented for very long hours, stinker beans will be produced. Uncontrolled fermentation and incomplete mucilage removal contribute to the bad quality of coffee (Oumer and Abate 2017). Therefore, pectinase enzyme can be used to enhance the removal of mucilage coat from coffee beans.

6.4.4 Tea Processing

Tea is obtained from the leaves of *Camellia sinensis* and the bud is classified based on the manufacturing process, with the black tea being comprised of fermentation and firing steps and the green tea being involved in dried process (Chandini et al. 2011). The beneficial effect for utilization of tea flavonoids has been connected to lower rates of chronic diseases, for example, cardiovascular ailment and cancer (de Koning Gans et al. 2010).

Pectinases are industrially produced in different kinds of fermentation process technology to be used as processing aids for extraction, clarification, and maceration objectives (Sharma et al. 2013; Kubra et al. 2018). Pectinase is a group of enzymes that hydrolyzes pectin to accelerate the fermentation process. Throughout the fermentation procedure, the tea catechins acting as oxidizing agents for the constituents as gallic acid are oxidized to form orthoquinones, which condensed to form theaflavins (TF). TF are responsible for the astringency characteristics to the taste of finished tea and are accountable for the briskness, brightness and quality of the liquor (Angayarkanni et al. 2002; Borse and Jagan 2012; Asil et al. 2012; Ghosh et al. 2012). Oxidation of gallic acid occurs to form epitheaflavic acids, then condensed with theaflavins to generate the polymeric thearubigins (TR) (Angayarkanni et al. 2002). The presence of TR is an indication of the intrinsic quality of color, flavor, body, and taste formation (Borse and Jagan 2012; Asil et al. 2012). The perfect fermentation process outcome is usually a genuine equilibrium between theaflavins and thearubigins. The volatile compounds are produced in mixture from the deterioration of lipid during a devastating step in small amount during tea leaves processing. These volatile compounds determine the final aroma during tea manufacturing. Rapid and accomplished maceration and complete aeration, therefore, are crucial conditions for excellent fermentation process (Angayarkanni et al. 2002). Table 6.8 shows the different strain producers for enzyme pectinase and its functions to improve the quality control for tea.

During the tea fermentation, pectinase enzymes from a microorganism act to rupture pectin present in the cell walls of tea leaves; however, the overuse of these enzymes can damage tea leaves, so a particular volume must be controlled during fermentation procedure. The enzyme also functions as an antifoaming agent by devastating pectin in the instant tea powders. The adjustment in the color of tea during the fermentation likewise brings about the improvement of the trademark smell (Praveen and Suneetha 2014). The role of enzymes in tea making and their application for maceration of tea leaves and improvement of feature and characteristics of final produce is presented in Table 6.8. The quality control parameters for tea studies were theaflavins (TF), thearubigins (TR), highly polymerized substances (HPS), total liquor color (TLC), dry matter content (DMC), and total soluble solids (TSS) (Angayarkanni et al. 2002; Murugesan et al. 2002; Asil et al. 2012; Borse and Jagan 2012).

Table 6.8 Different strains for pectinase production and their functions to improve tea quality

Strain used	Functions	Improves tea quality by increasing (%)	Enzyme involved
A. oryzae, A. wentii, A. tamari, A. japonicus, A. awamori, and Trichoderma koningii	Enhanced the tea quality	TF by 45	Mixed enzyme extract
		TR by 48	
		HPS by 33	
		TLC by 19	
Senthilkumar et al. (2000)		TSS by 3	
A. indicus	Enhanced the TF and TLC content	TF by 38.10	Purified pectinase enzyme
A. flavus		TF by 40	
A niveus		TF by 34.29	
Angayarkanni et al. (2002)		TLC by 12.18	
		TLC by 11.54	
		TLC by 11.22	
A. indicus	Enhanced the TF and TLC content	TF by 43.81	Mixtures of crude enzyme (pectinases, cellulases, hemicellulases, proteinases)
A. flavus		TF by 62.86	
A niveus		TF by 59.05	
Angayarkanni et al. (2002)		TLC by 18.19	
		TLC by 14.74	
		TLC by 14.10	
Aspergillus flavus, Aspergillus niger	Increase caffeine production	Caffeine in tea to 0.49%	Pectin lyase
(Pichia sp. NRRL Y-4810)	Increased black tea components	TF by 9.52	Pectinase
(Zygosaccharomyces sp. NRRL Y-4882)	Theaflavin (TF),	TR by 9.48	
(Acetobacter sp. NRRL B-2357)	Thearubigin (TR),	HPS by 9.65	
Murugesan et al. (2002)	Caffeine (CAF)	TLC by 3.29	
	Highly polymerized substances (HPS)	TSS by 37.64	
	Total liquor color (TLC)	CAF by 2.76	
	Total soluble solids (TSS)	DMC by 96.7	
	Dry matter content (DMC)		

6.4.5   Chocolate Processing

In cocoa fermentation, several microorganisms are necessary to generate the chocolate flavor. The microorganism is involved in the degradation of the cocoa pulp by liberating pectinase and standard grade one quality of cocoa beans is obtained with premium flavor (Schwan and Wheal 2004; Kubra et al. 2018). A study done by Bhumibhamon and Jinda (1997) observed the effects of soaking cocoa beans in 200 ppm pectinase enzyme and the quality of fermented cocoa beans with no slaty and mold beans under acceptable pH values of 5.49 and 5.92. The addition of pectin lyase in a concentration between 0.5% and 1.0% at a certain time during the fermentation process lowers the drying time of the cocoa bean and thus improve the chocolate quality (Gil et al. 2016).

6.4.6   Animal Feed Application

Other main industrial applications of the pectinase are in an animal feed processing industry. According to Menezes-Blackburn and Greiner (2015), feed becomes the highest expense in large-scale breeding of monogastric animals, accounting for more than 70% of the total production cost. The application of pectinase alone has accounted for 10% of globally produced industrial enzymes and continues to expand over time (Kubra et al. 2018). The increase in supplement animal diets with hydrolytic enzymes is also driven by the consumer awareness in ingredient quality and new legislation that prohibited any growth-promoting drugs and chemical substances to increase animal production (Walsh et al. 1993).

The animal feed industry uses enzymes that degrade crude fiber, starch, proteins, and phytates. These are eventually digested or excreted by the animal with no residual effect on the products such as meat or egg (Bedford and Partridge 2001). Currently, pectinase has been used as animal feed supplementation to enzymes to increase nutrient absorption other than phytases, xylanases, and β-glucanases (Kubra et al. 2018). The supplementation of feed with enzymes is also increasing the digestive effectiveness towards undiscriminating factors naturally present in feed ingredients (such as fiber), especially when the animal's digestive system is not fully efficient (Garg et al. 2016; Ojha et al. 2019). Hoondal et al. (2002) reported that pectinase used in animal feed helps in the reduction of feed viscosity as well as the animal feces.

According to Bedford (2018), even though the enzymes have been commercialized for over 40 years, current research is still in debate as to the use of single or multi-enzymes that suited the best rations. Meanwhile, until 2008, the use of feed enzymes in aquaculture is limited and even non-existent in ruminants (Barletta 2010). Jackson (2010) reported that some factors such as animal species, examination of stage of production, feed substrates, and the activity level of pectinase used are among the factors that need to be considered to increase the efficacy of the enzymes in combination with pectinase in animal feed.

6.5   Commercial Pectinases

Pectinase shows widespread commercial use, especially in the food, textile, and cosmeceutical industry. Its commercial application was first reported in 1930 for the preparation of fruit juices (Kertesz 1930). Nowadays, different suppliers worldwide produce this enzyme in commercial scale for different industries (Table 6.9).

6.6   Conclusion

Pectinases are an important group of enzymes that contribute to the degradation of pectin, which is a complex acidic polysaccharide present in the primary cell wall of the plant. Pectinolytic enzymes from fungal sources can be applied in various industrial sectors due to their efficient and stable production process. The significance of these enzymes for the development of environmentally friendly industrial processes has already been established. Further research on the enzyme production process to improve the process economy is still needed to increase the scope of application in new industries. Furthermore, research on enzyme engineering to increase stability and activity under different operation conditions is a matter of interest for different research group. These all together will help to increase market accessibility and open new niche area for industrial application of this important enzyme.

Table 6.9 Different commercial pectinases commonly used in industry

Product commercial name	Supplier
Rohapect® MA Plus	AB Enzymes (Rajamaki, Finland)
Chemos	CHEMOS GmbH (Altdorf, Germany)
Pectinase	West Bengal Chemical Industries Ltd.(Kolkata, India)
Pectinex® Ultra Mash	Novozymes (Bagsvaerd, Denmark)
Pectinase from A. niger	Tokyo Chemical Industry Co. Ltd. (Saitama, Japan)
Ly Peclyve PR®	Lyven (Pays de Caen Colombelles, France)
Polygalakturonase	Kanto Chemical Co., Inc. (Chuo-ku, Japan)
Klerzyme®	Wallerstein Co. (Colorado, USA)
Pectinex®	Schweizerische Ferment A.G. (Zug, Switzerland)
Panzym®	C.H. Boehringer Sohn AG & Co. (Ingelheim, Germany)
Macerozyme R-10	Apollo Scientific Ltd. (Stockport SK6 2QR, United Kingdom)
Pectase	Hubei Hongxin Ruiyu Fine Chemical Co., Ltd. (Shanghai, China)

Data modified from Garg et al. (2016)

Acknowledgments The authors would like to thank the Ministry of Education (Malaysia) and Universiti Teknologi Malaysia (UTM) for HICOE grant No. R.J130000.7851.4 J386.

References

Abas Z, Uzir H, Zahar M (2010) Effect of pH on the biotransformation of (*R*)-1-(4-bromo-phenyl)-ethanol by using *Aspergillus niger* as biocatalyst. J Appl Sci 10(24):3289–3294

Abdullah R, Farooq I, Kaleem A, Iqtedat M, Iftikhar T (2018) Pectinase production from *Aspergillus niger* IBT-7 using solid state fermentation. Bangladesh J Bot 47(3):473–478

Aguilar G, Huitrón C (1990) Constitutive exo-pectinase produced by *Aspergillus sp.* CH-Y-1043 on different carbon source. Biotechnol Lett 12(9):655–661

Ahmed I, Zia MA, Hussain MA, Akram Z, Naveed MT, Nowrouzi A (2016) Bioprocessing of citrus waste peel for induced pectinase production by *Aspergillus niger*; its purification and characterization. J Rad Res Appl Sci 9(2):148–154

Alimardani TP, Gainvors CA, Duchiron F (2011) Yeasts: an attractive source of pectinases-from gene expression to potential applications. A review. Process Biochem 46(8):1525–1537

Alkorta I, Garbisu C, Llama MJ, Serra JL (1998) Industrial applications of pectic enzymes: a review. Process Biochem 33(1):21–28

Alvarez S, Alvarez R, Riera F, Coca J (1998) Influence of depectinization on apple juice ultrafiltration. Colloids Surfaces Physicochem 138:377–382

Anand G, Yadav S, Yadav D (2017) Production, purification and biochemical characterization of an exo-polygalacturonase from *Aspergillus niger* MTC 478 suitable for clarification of orange juice. 3 Biotech 7(122):1–8

Angayarkanni J, Palaniswamy M, Murugesan S, Swaminathan K (2002) Improvement of tea leaves fermentation with *Aspergillus sp.* pectinase. J Biosci Bioeng 94:299–303

Arijit D, Sourav B, Naimisha R, Rajan S (2013) Improved production and purification of pectinase from *Streptomyces* sp. GHBA10 isolated from Valapattanam mangrove habitat, Kerala, India. Int Res 2(3):16–22

Arotupin DJ (2007) Effect of different carbon sources on the growth and polygalacturonase activity of *Aspergillus flavus* isolated from cropped soils. Res J Microbiol 2(4):362–368

Asgher M, Shaha WK, Muhammad B (2016) Optimization of lignocellulolytic enzyme production by *Pleurotus eryngii* WC 888 utilizing agro-industrial residues and bio-ethanol production. Rom Biotech Lett 21(1):11133–11143

Asil MH, Rabiei B, Ansari RH (2012) Optimal fermentation time and temperature to improve biochemical composition and sensory characteristics of black tea. Aust J Crop Sci 6(3):550

Barbosa OCO, Berenguer-Murcia R, Torres RC, Rodrigues R, Fernandez L (2015) Strategies for the one-step immobilization-purification of enzymes as industrial biocatalysts. Biotechnol Adv 33:435–456

Barletta A (2010) Thermostability of feed enzymes and their practical application in the feed mill. In: Bedford MR, Partridge GJ (eds) Sci Food Agric. CABI, Cambridge, pp 1–11

Bayindirli A (2010) Enzymes in fruit and vegetable processing. In: Int J Chem Eng. CRC Press, Hoboken

Bedford MR (2018) The evolution and application of enzymes in the animal feed industry: the role of data interpretation. Br Poult Sci 59(5):486–493

Bedford MR, Partridge GG (2001) Enzymes in farm animal nutrition. J Anim Physiol Anim Nutr 85:333–334

Bhumibhamon O, Jinda J (1997) Effect of enzyme pectinase on natural cocoa fermentation. Kasetsart J (Nat Sci) 31:206–212

Biz A, Finkler ATJ, Pitol LO, Medina BS, Krieger N, Mitchell DA (2016) Production of pectinases by solid-state fermentation of a mixture of citrus waste and sugarcane bagasse in a pilot-scale packed-bed bioreactor. Biochem Eng J 111:54–62

Blanco P, Sieiro C, Villa TG (1999) Production of pectic enzymes in yeasts. FEMS Microbiol Lett 175(1):1–9

Borse BB, Jagan MRL (2012) Novel bio-chemical profiling of Indian black teas with reference to quality parameters. J Bioequivalence Bioavailab 14:1–16

Bussink J, Buxton P, Fraaye B, Graaff H, Visser J (1992) The polygalacturonases of *Aspergillus niger* are encoded by a family of diverged genes. Eur J Biochem 208(1):83–90

Celestino SMC, De SM, Medrano FJ, De MV, Ferreira EX (2006) Purification and characterization of a novel pectinase from *Acrophialophora nainiana* with emphasis on its physicochemical properties. J Biotechnol 123(1):33–42

Chandini SK, Jaganmohan RL, Gowthaman MK, Haware DJ, Subramanian (2011) Enzymatic treatment to improve the quality of black tea extracts. Food Chem 127:1039–1045

Chandrasekaran M, Basheer SM, Chellappan S, Krishna JG, Beena PS (2015) Enzymes in food and beverage production: an overview. In: Int J Chem Eng. CRC Press, Hoboken, pp 133–154

Channe S, Sitewale G (1995) Pectinase production by *Sclerotium rolfsii*: effect of culture conditions. Folia Microbiol 40(I):111–117

Contreras EJC, Voget CE (2004) Purification and partial characterization of an acid polygalacturonase from *Aspergillus kawachii*. J Biotechnol 110:21–28

de Koning Gans JM, Uiterwaal CS, van der Schouw YT, Boer JM, Grobbee DE, Verschuren WM, Beulens JW (2010) Tea and coffee consumption and cardiovascular morbidity and mortality. Arterioscler Thromb Vasc Biol 30(8):1665–1671

de Vries RP, Visser JAAP (2001) Aspergillus enzymes involved in degradation of plant cell wall polysaccharides. Microbiol Mol Biol Rev 65(4):497–522

de Vries J, Wackernagel W (2002) Integration of foreign DNA during natural transformation of *Acinetobacter* sp. by homology-facilitated illegitimate recombination. Proc Natl Acad Sci 99(4):2094–2099

de Vries J, van de Vondervoort P, Hendriks L, Van de Belt M, Visser J (2002) Regulation of the alpha-glucuronidase-encoding gene (aguA) from *Aspergillus niger*. Mol Gen Genomics 268(1):96–102

Debing J, Peijun L, Stagnitti F, Xianzhe X, Li L (2006) Pectinase production by solid fermentation from *Aspergillus niger* by a new prescription experiment. Ecotoxicol Environ Saf 64:244–250

Deng LZ, Pan Z, Zhang Q, Liu ZL, Zhang Y, Meng JS, Gao ZJ, Xiao HW (2019) Effects of ripening stage on physicochemical properties, drying kinetics, pectin polysaccharides contents and nanostructure of apricots. Carbohyd Polym 2019:114980

Doughari JH, Onyebarachi GC (2019) Production, purification and characterization of polygalacturonase from *Aspergillus flavus* grown on orange peel. Appl Microbiol 4:159

El-Enshasy A, Kleine J, Rinas U (2006) Agitation effects on morphology and protein productive fractions of filamentous and pelleted growth forms of recombinant *Aspergillus niger*. Process Biochem 41:2103–2112

El-Enshasy HA, Elsayed EA, Suhaimi N, Abd MR, Esawy M (2018) Bioprocess optimization for pectinase production using *Aspergillus niger* in a submerged cultivation system. BMC Biotechnol 18:71

Esawy MA, Gamal AA, Kamel Z, Ismail AS, Abdel-Fattah AF (2013) Evaluation of free and immobilized *Aspergillus niger* NRC1ami pectinase applicable in industrial processes. Carbohyd Polym 92:1463–1469

Farid A, El-Enshasy A, El-Diwany I, El-Sayed A (2000) Optimization of the cultivation medium for natamycin production by *Streptomyces natalensis*. J Basic Microbiol 40(3):157–166

Favela-Torres E, Volke-Sepúlveda T, Viniegra-González G (2006) Production of hydrolytic depolymerising pectinases. Food Technol Biotechnol 44(2):221–227

Fernández-González M, Ubeda JF, Vasudevan TG, Otero RC, Briones AI (2004) Evaluation of polygalacturonase activity in *Saccharomyces cerevisiae* wine strains. FEMS Microbiol Lett 237:261–266

Gao Y, Fangel JU, Willats WG, Vivier MA, Moore JP (2016) Dissecting the polysaccharide-rich grape cell wall matrix using recombinant pectinases during winemaking. Carbohyd Polym 152:510–519

Garg G, Singh A, Kaur A, Singh R, Kaur J, Mahajan R (2016) Microbial pectinases: an ecofriendly tool of nature for industries. 3 Biotech 6:47

Georgiou G, Segatori L (2005) Preparative expression of secreted protein in bacteria: status report and future prospects. Curr Opin Biotech 16:538–545

Ghosh A, Tamuly P, Bhattacharyya N, Tudu B, Gogoi N, Bandyopadhyay R (2012) Estimation of theaflavin content in black tea using electronic tongue. J Food Eng 110(1):71–79

Gil M, Orrego F, Cadena E, Alegria R, Londono-Londono J (2016) Effect of pectin lyase enzyme on fermentation and drying of cocoa (*Theobroma cacao L.*): an alternative to improve raw material in the industry of chocolate. Int J Food Sci Nutr 7:215–226

Global Pectinase Market Research Report (2017). http://www.marketresearchstore.com/report/global-pectinase-marketresearch-report2017-190713

Gonçalves DB, Teixeira JA, Bazzolli DMS, Queiroz MVD, Araújo EFD (2012) Use of response surface methodology to optimize production of pectinases by recombinant *Penicillium griseoroseum* T20. Biocatal Agric Biotechnol 1(2):140–146

Gummadi N, Panda T (2003) Purification and biochemical properties of microbial pectinases-a review. Process Biochem 38:987–996

Guo F, Li X, Zhao J, Li G, Gao P, Han X (2019) Optimizing culture conditions by statistical approach to enhance production of pectinase from *Bacillus sp.* Y1. BioMed Res Int 2019:8146948

Hadj-Taieb N, Ayadi M, Khlif M, Mrad K, Hassairi I, Gargouri A (2006) Fermenter production of pectinases on gruel, a local by-product and their use in olive oil extraction. Enzym Microb Technol 39(5):1072–1076

Harmsen M, Someren K, Visser J, Kustersvansomeren A, Visser J (1990) Cloning and expression of a second *Aspergillus niger* pectin lyase gene. Curr Genet 593(2):161–166

Hoondal G, Tiwari R, Tewari R, Dahiya N, Beg Q (2002) Microbial alkaline pectinases and their industrial applications: a review. Appl Microbiol Biotechnol 59(4–5):409–418

Hyson DA (2015) A review and critical analysis of the scientific literature related to 100% fruit juice and human health. Adv Nutr 6:37–51

Iwashita K (2002) Review: recent studies of protein secretion by filamentous fungi. J Biosci Bioeng 94(6):530–535

Jackson ME (2010) Mannanase, alpha-galactosidase and pectinase. In: Bedford MR, Partridge G (eds) J Sci Food Agric. CABI, Wallingford, pp 54–84

Jahic M, Rotticci-Mulder C, Martinelle M, Hult K, Enfors O (2002) Modelling of growth and energy metabolism of *Pichia pastoris* producing a fusion protein. Bioprocess Biosyst Eng 24:385–393

Javed R, Nawaz A, Munir M, Hanif MU, Mukhtar H, Ul Haq I, Abdullah I (2018) Extraction, purification and industrial applications of pectinase: a review. J Biotechnol Biores 1:1. JBB.000503.2018

Jayani S, Saxena S, Gupta R (2005) Microbial pectinolytic enzymes: a review. Process Biochem 40(9):2931–2944

Joshi K, Parmar M, Rana N (2011) Purification and characterization of pectinase produced from apple pomace and evaluation of its efficacy in fruit juice extraction and clarification. Indian J Nat Prod Resour 2(2):189–197

Kashyap DR, Chandra S, Kaul A, Tewari R (2000) Production, purification and characterization of pectinase from a *Bacillus sp.* DT7. World J Microbiol Biotechnol 16(3):277–282

Kashyap DR, Vohra PK, Chopra S, Tewari R (2001) Applications of pectinases in the commercial sector: a review. Bioresour Technol 77(3):215–227

Kaur A, Mahajan R, Singh A, Garg G, Sharma J (2010) Application of cellulase-free xylano-pectinolytic enzymes from the same bacterial isolate in biobleaching of kraft pulp. Bioresour Technol 101(23):9150–9155

Kavanagh K (2005) Fungi biology and applications. John Wiley and Sons, Ltd, London

Kavuthodi B, Sebastian D (2018) Review on bacterial production of alkaline pectinase with special emphasis on *Bacillus* species. Biosci Biotechnol 11(1):18–30

Kertesz Z (1930) A new method for enzymic clarification of unfermented apple juice. US patent no. 1.932.833, New York State Agricultural Experimentation Station (Geneva) Bull. No. 689

Khairnar Y, Krishna VK, Boraste A, Gupta N, Trivedi S, Patil P, Gupta G, Gupta M, Jhadav A, Mujapara A, Joshi B, Mishra D (2009) Study of pectinase production in submerged fermentation using different strains of *Aspergillus niger*. Int J Microbiol Res 1(2):13

Khatri BP, Bhattarai T, Shrestha S, Maharjan J (2015) Alkali thermostable pectinase enzyme from *Aspergillus niger* strain MCAS2 isolated from Manaslu Conservation Area, Gorkha, Nepal. SpringerPlus 4:488

Khosravi C, Benocci T, Battaglia E, Benoit I, de Vries RP (2015) Sugar catabolism in Aspergillus and other fungi related to the utilization of plant biomass. Adv Appl Microbiol 90:1–28

Knoll A, Bartsch S, Husemann B, Engel P, Schroer K, Ribeiro B, Stöckmann C, Seletzky J, Büchs J (2007) High cell density cultivation of recombinant yeasts and bacteria under non-pressurized and pressurized conditions in stirred tank bioreactors. J Biotechnol 132(2):167–179

Kubra KT, Ali S, Walait M, Sudus H (2018) Potential applications of pectinase in food, agricultural and environmental sectors. J Pharm Chem Biol Sci 6(2):23–24

Kuivanen J, Mojzita D, Wang Y, Hilditch S, Penttilä M, Richard P, Wiebe MG (2012) Engineering filamentous fungi for conversion of D-galacturonic acid to L-galactonic acid. Appl Environ Microbiol 78(24):8676–8683

Kuorelahti S, Jouhten P, Maaheimo H, Penttilä M, Richard P (2006) L-galactonate dehydratase is part of the fungal path for D-galacturonic acid catabolism. Mol Microbiol 61(4):1060–1068

Leyva SM, Mounier J, Valence F, Coton M, Thierry A, Coton E (2017) Antifungal microbial agents for food biopreservation-a review. Microorganisms 5(3):37

Li S, Yang X, Yang S, Zhu M, Wang X (2012) Technology prospecting on enzyme: application, marketing and engineering. Comput Struct Biotechnol J Journal 2(3):1–11

Li X, Yu C, Guo Y, Bian Z, Si J, Yang L, Chen Y, Ren X, Jiang G, Chen J, Chen Z, Lv J, Li L (2017) Tea consumption and risk of ischaemic heart disease. Heart 103(10):783–789

Martens-Uzunova ES, Schaap PJ (2008) An evolutionary conserved D-galacturonic acid metabolic pathway operates across filamentous fungi capable of pectin degradation. Fungal Genet Biol 45:1449–1457

Martens-uzunova ES, Schaap PJ (2009) Assessment of the pectin degrading enzyme network of *Aspergillus niger* by functional genomics. Fungal Genet Biol 46(Suppl 1):S170–S179

Martins ES, Silva D, Leite RS, Gomes E (2007) Purification and characterization of polygalacturonase produced by thermophilic *Thermoascus aurantiacus* CBMAI-756 in submerged fermentation. Antonie Van Leeuwenhoek 91(3):291–299

Mehmood T, Saman T, Irfan M, Anwar F, Ikram MS, Tabassam Q (2019) Pectinase production from *Schizophyllum commune* through central composite design using citrus waste and its immobilization for industrial exploitation. Waste Biomass Valoriz 10(9):2527–2536

Menezes-Blackburn D, Greiner R (2015) Enzymes used in animal feed: leading technologies and forthcoming developments. In: Cirillo G, Spizzirri UG, Iemma F (eds) Reactive and functional polymers. Wiley, Calabria, pp 47–73

Merín MG, Morata de Ambrosini VI (2015) Highly cold-active pectinases under wine-like conditions from non-*Saccharomyces* yeasts for enzymatic production during winemaking. Lett Appl Microbiol 60:467–474

Miyazaki T, Matsumoto Y, Matsuda K, Kurakata Y, Matsuo I, Ito Y, Nishikawa A, Tonozuka T (2011) Heterologous expression and characterization of processing α-glucosidase i from *Aspergillus brasiliensis* ATCC 9642. Glycoconj J 28:563–571

Mojsov K, Andronikov D, Janevski A, Jordeva S, Zhezhova S (2015) Enzymes and wine–the enhanced quality and yield. Polym Adv Technol 4:94–100

Murugesan GS, Angayarkanni J, Swaminathan K (2002) Effect of tea fungal enzymes on the quality of black tea. Food Chem 79:411–417

Naidu G, Panda T (1998) Application of response surface methodology to evaluate some aspects on stability of pectolytic enzymes from *Aspergillus niger*. Biochem Eng J 2(1):71–77

Nevalainen H, Te'o J, Bergquist L (2005) Heterologous protein expression in filamentous fungi. Trends Biotechnol 23(9):468–474

Ngo LTA, Pham TL, Le VVM (2008) Purification of Endopolygalacturonase from submerged culture of *Aspergillus awamori* L1 using a two-step procedure: enzyme precipitation and gel filtration. Int Food Res J 15(2):135–140

Nighojkar A, Patidar M, Nighojkar S (2019) Pectinases: production and applications for fruit juice beverages. Sustainability 2:235–273

Ojha BK, Singh PK, Shrivastava N (2019) Enzymes in the animal feed industry. In: Kuddus M (ed) Enzyme and microbial technology. Academic Press, Oxford, pp 93–109

Okinji RE, Itakorode BO, Ovumedia JO, Adedeji OS (2019) Purification and biochemical characterization of pectinase produced by *Aspergillus fumigatus* isolated from soil of decomposing plant materials. J Appl Biol Biotechnol 7(03):1–8

Oumer OJ (2017) Pectinase: substrate, production and their biotechnological applications. Int J Environ Agric Biotechnol 2:1007–1014

Oumer OJ, Abate D (2017) Characterization of pectinase from *Bacillus subtilis* strain Btk 27 and its potential application in removal of mucilage from coffee beans. Enzyme Res 2017:1–7

Papagianni M (2004) Review: fungal morphology and metabolite production in submerged mycelia processes. Biotechnol Adv 22:189–259

Pařenicová L, Benen E, Kester M, Visser J (1998) PgaE encodes a fourth member of the endopolygalacturonase gene family from *Aspergillus niger*. Eur J Biochem 251(1–2):72–80

Pařenicova L, Benen E, Kester M, Visser J (2000) PgaA and pgaB encode two constitutively expressed endopolygalacturonases of *Aspergillus niger*. Biochem J 345:637–644

Pasha KM, Anuradha P, Subbarao D (2013) Applications of pectinases in industrial sector biotechnological applications of microbial pectinases. Int J Pure Appl Sci Technol 16(1):89–95

Patel A, Singhania R, Pandey A (2016) Novel enzymatic processes applied to the food industry. Curr Opin Food Sci 7:64–72

Patidar MK, Nighojkar A, Nighojkar S, Kumar A (2017) Purification and characterization of polygalacturonase produced by *Aspergillus niger* AN07 in Solid State Fermentation. Can J Microbiol 1(1):11

Pedrolli B, Monteiro C, Gomez E, Carmona C (2009) Pectin and pectinases: production, characterization and industrial application of microbial pectinolytic enzymes. Open Biotechnol J 3(1):9–18

Poletto P, Borsói C, Zeni M, Moura M (2015) Downstream processing of pectinase produced by *Aspergillus niger* in solid state cultivation and its application to fruit juices clarification. Food Sci Technol 35(2):391–397

Porro D, Gasser B, Fossati T, Maurer M, Branduardi P, Sauer M, Mattanovich D (2011) Production of recombinant proteins and metabolites in yeasts. Appl Microbiol Biotechnol 89(4):939–948

Praveen KG, Suneetha V (2014) A cocktail enzyme-pectinase from fruit industrial dump sites: a review. Res J Pharm Biol Chem Sci 5(2):1252–1258

Punt J, Biezen V, Conesa A, Albers A, Mangnus J, Hondel D (2002) Filamentous fungi as cell factories for heterologous protein production. Trends Biotechnol 20(5):200–2007

Rai M, Padh H (2001) Expression systems for production of heterologous protein. Curr Sci 80(9):1121–1128

Rai P, Majumdar GC, Dasgupta SDES, De S (2004) Optimizing pectinase usage in pretreatment of mosambi juice for clarification by response surface methodology. J Food Eng 64:397–403

Ramadan MF (2019) Enzymes in fruit juice processing. In: Enzymes in food biotechnology. Academic Press, Oxford, pp 45–59

Richard P, Hilditch S (2009) D-Galacturonic acid catabolism in microorganisms and its biotechno-
logical relevance. Appl Microbiol Biotechnol 82:597–604
Said S, Fonseca V, Siéssera V (1991) Pectinase production by *Penicillium frequentans*. World J
Microbiol Biotechnol 7:607–608
Sakai Y, Goh TK, Tani Y (1993) High-frequency transformation of a methylotrophic yeast,
Candida boidinii, with autonomously replicating plasmids which are also functional in
Saccharomyces cerevisiae. J Bacteriol 175(11):3556–3562
Sandri IG, Moura da Silveira M (2018) Production and application of pectinases from *Aspergillus
niger* obtained in solid state cultivation. Beverages 4(3):48
Sandri IG, Fontana RC, Barfknecht DM, da Silveira MM (2011) Clarification of fruit juices by
fungal pectinases. LWT-Food Sci Technol 44(10):2217–2222
Sarrouh B, Santos M, Miyoshi A, Dias R, Azevedo V (2012) Up-to-date insight on industrial
enzymes applications and global market. J Bioprocess Biotechnol 2:1–10. https://doi.
org/10.4172/2155-9821.S4-002
Schmidt R (2004) Recombinant expression systems in the pharmaceutical industry. Appl Microbiol
Biotechnol 65:363–372
Schuster E, Dunn-Coleman N, Frisvad JC, Van DPW (2002) On the safety of *Aspergillus niger*-a
review. Appl Microbiol Biotechnol 59(4–5):426–435
Schwan RF, Wheal AE (2004) The microbiology of cocoa fermentation and its role in chocolate
quality. Critical Rev Food Sci Nutr 44:205–221
Sealy-Lewis M, Fairhurst V (1992) An NADP⁺ –dependent glycerol dehydrogenase in *Aspergillus
nidulans* is inducible by D-galacturonate. Curr Genet 22:293–296
Semenova MV, Grishutin SG, Gusakov AV, Okunev ON, Sinitsyn AP (2003) Isolation and properties of pectinases from the fungus *Aspergillus japonicus*. Biochem Mosc 68(5):559–569
Senthilkumar RS, Swaminathan K, Marimuthu S, Rajkumar R (2000) Microbial enzymes for processing of tea leaf. In: Muraleedharan N, Kumar RR (eds) Recent adv plant crops res. Allied
publishers limited, Chennai, pp 265–269
Sharma N, Rathore M, Sharma M (2013) Microbial pectinase: sources, characterization and applications. Rev Environ Sci Biol 12(1):45–60
Sharma HP, Patel H, Sugandha (2017) Enzymatic added extraction and clarification of fruit juices-
a review. Crit Rev Food Sci Nutr 57(6):1215–1227
Singh JJ (2018) Pectinase: a useful tool in fruit processing industries. Int J Food Sci Nutr
2(3):125–129
Singh G, Kaur S, Khatri M, Arya SK (2019) Biobleaching for pulp and paper industry in India:
emerging enzyme technology. Biocatal Agric Biotechnol 17:558–565
Soares I, Távora Z, Barcelos RP, Baroni S (2012) Microorganism-produced enzymes in the food
industry. In: Food industry, scientific, health and social aspects of the food industry. IntechOpen,
New York, pp 83–94
Solis-Pereyra S, Favela-Torres E, Viniegra-Gonzalez G, Gutierrez-Rojas M (1993) Effect of different carbon sources on the synthesis of pectinases by *Aspergillus niger* in submerged and solid
state fermentations. Appl Microbiol Biotechnol 39:36–41
Suhaimi N, Ramli S, Malek RA, Aziz R, Zalina N, Leng OM, Esawy M, Gamal A, El-enshasy H
(2016) Research article optimization of pectinase production by *Aspergillus niger* using orange
pectin based medium. J Chem Pharm Res 8(2):259–268
Tapre AR, Jain RK (2014) Pectinases: enzymes for fruit processing industry. Int Food Res J
21(2):447–453
Toushik SH, Lee KT, Lee JS, Kim KS (2017) Functional applications of lignocellulolytic enzymes
in the fruit and vegetable processing industries. J Food Sci 82:585–593
Tripathi NK, Sathyaseelan K, Jana AM, Rao PVL (2009) High yield production of heterologous
proteins with *Escherichia coli*. Defence Sci J 59(2):137–146
Verma H, Narnoliya LK, Jadaun JS (2018) Pectinase: a useful tool in fruit processing industries.
Int J Food Sci Nutr 5(5):555673. https://doi.org/10.19080/NFSIJ.2018.05.555673

Voragen AGJ, Pilnik W, Thibault JF, Axelos MAV, Renard CMGC (1995) Pectins. In: Stephen AM (ed) Food polysaccharides and their applications. Marcel Dekker, Inc., New York, pp 287–369

Wainwright M (1992) An introduction to fungal biotechnology. University of Sheffield, UK. John Wiley and Son, Sheffield

Walsh GA, Ronan F, Power D, Headon R (1993) Enzymes in the animal-feed industry. Trends Biotechnol 11(10):424–430

Wang L, Ridgway D, Gu T, Moo-Young M (2005) Bioprocessing strategies to improve heterologous protein production in filamentous fungal fermentations. Biotechnol Adv 23:115–129

Westers L, Westers H, Quax WJ (2004) *Bacillus subtilis* as cell factory for pharmaceutical proteins: a biotechnological approach to optimize the host organism. Biochim Biophys Acta 1694:299–300

Yadav S, Yadav PK, Yadav D, Yadav KDS (2008) Purification and characterization of an alkaline pectin lyase from *Aspergillus flavus*. Process Biochem 43(5):547–552

Yang J, Luo H, Li J, Wang K, Cheng H, Bai Y, Yuan T, Fan Y, Yao B (2011) Cloning, expression and characterization of an acidic endopolygalacturonase from *Bispora sp.* MEY-1 and its potential application in juice clarification. Process Biochem 46(1):272–277

Yuan P, Meng K, Huang H, Shi P, Luo H, Yang P, Yao B (2011) A novel acidic and low-temperature-active endo-polygalacturonase from *Penicillium sp.* CGMCC 1669 with potential for application in apple juice clarification. Food Chem 129(4):1369–1375

Zakharova A, Awan S, Nami F, Gotfredsen C, Madsen R, Clausen M (2018) Synthesis of two tetrasaccharide pentenyl glycosides related to the pectic rhamnogalacturonan i polysaccharide. Molecules 23(2):327

Zhang L, Thiewes H, van Kan JA (2011) The D-galacturonic acid catabolic pathway in *Botrytis cinerea*. Fungal Genet Biol 48(10):990–997

Chapter 7
Production of Polyunsaturated Fatty Acids by Fungal Biofactories and Their Application in Food Industries

Rishibha Gupta and Smriti Gaur

Abbreviations

ACC	acetyl-CoA carboxylase
ACL	ATP-citrate lyase
ALA	alpha-linoleic acid
ARA	arachidonic acid
C/N	carbon/nitrogen ratio
DHA	docosahexaenoic acid
EPA	eicosapentaenoic acid
GLA	gamma (γ)-linoleic acid
GRAS	generally regarded as safe
LA	linoleic acid
PUFA	polyunsaturated fatty acid
TCA	tricarboxylic acid

7.1 Introduction

In the early twentieth century, the role of lipids as an essential component in our diet was largely unknown. They were considered more as non-essential caloric content of food that can be replaced by increasing the carbohydrate content of the food. However, the discovery of linoleic acid as an essential fatty acid in the diet of rats by George and Mildred Burr marked the beginning of a new era of lipid research. It was found that rats fed with fat-deficient diet developed certain diseases and skin disorders which were cured by supplementation of linoleic acid in their diet. The latter studies soon began to highlight the significance of linoleic and other fatty

R. Gupta · S. Gaur (✉)
Department of Biotechnology, Jaypee Institute of Information Technology, Noida, India
e-mail: smriti.gaur@jiit.ac.in

© The Author(s), under exclusive license to Springer Nature
Switzerland AG 2021
X. Dai et al. (eds.), *Fungi in Sustainable Food Production*, Fungal Biology,
https://doi.org/10.1007/978-3-030-64406-2_7

acids in human and other animals (Spector and Kim 2015; Smith and Mukhopadhyay 2012). In this quest of lipids with nutritional value, polyunsaturated fatty acids (PUFAs) deserve a special mention.

PUFAs are a class of fatty acids characterized by the presence of long carbon chains with two or more double bonds. They are further broadly classified as ω-3, ω-6 and ω-9 on the basis of location of double bond with respect to terminal methyl group (Srianta et al. 2010; Kothri et al. 2020). The biosynthesis of PUFA occurs by an extension of saturated fatty acid pathway to produce linoleic acid, a precursor of ω-6 and ω-3 series of PUFA (Gill and Valivety 1997). A series of sequential desaturation and elongation steps occurs which introduces double bond at specific locations and adds two carbon atoms to the existing fatty acyl chain (Kim et al. 2014) as shown in Fig. 7.1. PUFAs are formed by a special class of desaturases called unsaturated fatty acid desaturases, and their type, activity and expression level are what determines the type of PUFA profile a cell is capable of generating (Czumaj and Śledziński 2020). The first microbial oil rich in γ-linolenic acid (GLA) was commercially produced by J&E Sturge (UK) using filamentous fungi *Mucor*

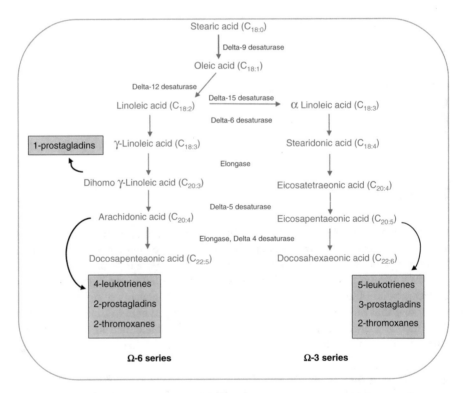

Fig. 7.1 Bioconversion of polyunsaturated fatty acids via aerobic desaturase/elongase enzymes with their potential bioactive role in mammals. (Adapted from Gill and Valivety 1997; Ochsenreither et al. 2016)

circinelloides from 1985 to 1990. Earlier, GLA was mainly obtained from evening primrose seeds making the oil expensive, but this new technology brought down the prices to a great extent (Ratledge 2013).

PUFAs have diverse roles in our body ranging from gene regulation to production of metabolites that have important biological roles. They act as a precursor to a number of eicosanoids and docosanoids involved in inflammatory responses of our body. Arachidonic acid (ARA) acts as a precursor for eicosanoids with pro-inflammatory roles, while ω-6 series eicosapentaenoic acid (EPA) and docosahexaenoic acid (DHA) act as precursors for the synthesis of anti-inflammatory compounds (Czumaj and Śledziński 2020). This distinction in the role of the two omega families is what makes the ratio of ω-6 and ω-3 an important factor for cardiovascular health and inflammatory responses (Sakuradani et al. 2009).

7.2   Oleaginous Fungi

Oleaginousity is a property of certain microorganisms to accumulate high amount of lipids in their cells usually in the form of triacylglycerol (Laoteng et al. 2011). Generally, microbes that store 20% or more lipids with respect to their cell dry weight are regarded as oleaginous. Some of these species are also able to synthesize long-chain unsaturated fatty acids which make them an attractive candidate for commercial PUFA production in large quantities (Khot et al. 2020).

In the kingdom fungi, PUFA is mainly produced by oleaginous moulds. Several genera including *Mortierella*, *Trichoderma*, *Pythium* and *Aspergillus* are able to produce essential PUFAs (Verma et al. 2020). Yeasts have been exploited in food and beverage industries since 6000 BC. One of the earliest yeast species known to mankind *is Saccharomyces cerevisiae* (McNeil and Stuart 2018). Therefore, their commercial exploitation in production of PUFA is highly sought-after. Oleaginous yeasts are well studied and some species are even reported to accumulate up to 80% lipids w/w of their cell dry weight. Common genera of oleaginous yeasts include *Rhodosporidium*, *Cryptococcus*, *Candida*, *Yarrowia*, *Trichosporon* and *Lipomyces* (Patel et al. 2020). However, oleaginous strains of yeasts are unable to produce PUFA naturally. Therefore, genetic manipulation techniques have been employed in them to produce economically important PUFA (Bellou et al. 2016). Since these species are oleaginous, endogenous saturated/monounsaturated fatty acids synthesized by these species can be directed towards a high level of PUFA synthesis. One of the models of oleaginous yeast for PUFA production is *Yarrowia lipolytica*. It is a GRAS fungus that can accumulate up to 50% of its cell dry weight as lipids and is known for its ability to utilize hydrophobic substrates. DuPont, a US-based company, has commercially exploited this species for EPA production by genetically engineering a strain that overexpresses 30 genes mainly associated with fatty acid desaturation and elongation and downregulating 4 genes involved in lipid degradation. Production of EPA as high as 56.6% of total fatty acid was reported in these strains (Gemperlein et al. 2019). Other than *Y. lipolytica*, *Lipomyces starkeyi* and

Rhodosporidium toruloides are also emerging as promising candidates for PUFA production mainly owing to their ability to utilize a range of agro-industrial wastes as substrates and their high level of lipid accumulation, thus enabling a more economic and eco-friendly production of PUFA. However, genetic engineering in these species is still a challenging endeavour as not much is known about their cellular regulatory mechanisms. However, many emerging technologies such as CRISPR/ Cas9 can be exploited in future for potential manipulation of these species (McNeil and Stuart 2018; Park et al. 2018).

7.2.1 Induction of Lipogenesis in Oleaginous Fungi

Lipogenesis in oleaginous fungi is generally achieved by cultivating these organisms in a medium containing excess carbon with limiting amounts of any essential nutrient such as vitamin or minerals. However, the most commonly employed strategy is nitrogen limitation. Besides, culture conditions such as temperature, pH and moisture content also directly account for the amount of lipid that can be accumulated (Bellou et al. 2016). In oleaginous yeast, the activity of NAD^+ isocitrate dehydrogenase is absolutely dependent on AMP. However, this dependency is not general in all oleaginous fungi as NAD^+ isocitrate dehydrogenases of *Mortierella alpina* and *Mucor circinelloides* are not completely ATP dependent (Laoteng et al. 2011). Under nitrogen stress conditions, AMP is converted to ammonium ions in the cell, and as a result NAD^+ isocitrate dehydrogenase activity is reduced. NAD^+ isocitrate dehydrogenase catalyzes the conversion of isocitrate to α-ketoglutaric acid, and as a result of its reduced activity, an increased amount of isocitrate accumulates in the mitochondria. Isocitrate exists in equilibrium with citrate in mitochondria mediated via aconitase enzyme. The accumulated citrate is then transported out of mitochondrial membrane via a citrate-malate shunt and converted to acetyl-CoA. The malate inside the mitochondria is subsequently converted to pyruvate via malic enzyme generating NADPH which can later be utilized for fatty acid synthesis. The ability to convert citrate in cytoplasm to acetyl-CoA by ATP-citrate lyase (ACL) is often regarded as a marker of oleaginous fungi. The acetyl-CoA is then converted to malonyl-CoA by the action acetyl-CoA carboxylase (ACC) enzyme. ACC activity is the first committed step towards fatty acid synthesis in yeast and fungi (Akpinar-Bayizit 2014; Laoteng et al. 2011). A diagrammatic representation of the above process is given in Fig. 7.2. Lipid biosynthesis is an energy-extensive process and requires huge amount of NADPH. Important sources of NADPH in oleaginous fungi include malic enzyme, pentose phosphate pathway, folate metabolism and isocitrate dehydrogenase in tricarboxylic acid (TCA) cycle (Wang et al. 2020).

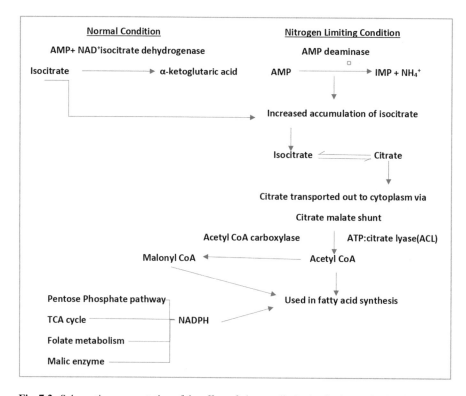

Fig. 7.2 Schematic representation of the effect of nitrogen limitation in the synthesis of fatty acids in oleaginous fungi

7.3   Production of PUFA from Fungal Cells

The first step in the production of commercial-level PUFA is identification of a robust fungal strain that is capable of accumulating sufficient amount of oil in its cell. The current advances in biotechnology enable high-throughput screening of fungal strains with a range of substrate and growth conditions in microtitre plates (Kosa et al. 2018). However, to produce a sustainable PUFA-producing cell biofactory, the metabolic pathway of oil synthesis in oleaginous cells is often redesigned to account for some of the following reasons:

- To increase the amount and type of desired PUFA
- To enable the fungal cells to utilize low-cost substrates
- To ease the downstream processing steps of lipid purification from these cells

Some of the common genetic engineering approaches involved in metabolic engineering of oleaginous fungi are discussed further.

Increasing NADPH Concentration in Cells
Since lipogenesis is an energy-consuming process, increasing NADPH concentration in cells can potentially increase lipid synthesis. Therefore, targeting NADPH-generating pathways can be viewed as a potential boost to PUFA accumulation. Two NADPH-generating enzymes of pentose phosphate pathway (PPP) and malic enzyme were overexpressed in native PUFA producers *Mortierella alpina*, *Mucor circinelloides* and *Aurantiochytrium* sp., and it was revealed that overexpression of malic enzyme increased desaturation of fatty acids, while PPP enzymes had a more significant effect on biosynthesis of fatty acids (Ji and Huang 2019).

Regulating the Activity of Desaturases and Elongases
Tailor-made PUFAs can be produced from a given microorganism by genetic reconstruction of the biosynthetic pathways involved in PUFA synthesis. The most common targets are desaturases and elongases. *Mucor circinelloides*, a filamentous fungus, had been mainly studied for its GLA production. However, a study by Khan et al. exploited the potential of *Mucor circinelloides* in the production of DGLA by the introduction of foreign delta-6 elongase enzyme coding gene from *Mortierella alpine* into it. Though the production was low (5.72%), the ease of purification of DGLA from other PUFA in this strain makes it a promising candidate for further exploration for commercial DGLA production (Khan et al. 2019). Overexpression of delta-12 fatty acid desaturase in *Rhodosporidium toruloides* has shown a fivefold increase in linoleic acid titre of these cells (Park et al. 2018).

Metabolite Regulation
Acetyl-CoA and citrate are important precursors of fatty acid synthesis. It has been shown in *Y. lipolytica* that deletion of PDH1 gene can increase citrate accumulation in cytoplasm irrespective of nitrogen-limiting conditions. Similarly, other active metabolites involved in fatty acid synthesis can also be increased in these cells by genetic manipulation (Bellou et al. 2016).

7.4 Large-Scale Cultivation of PUFA-Producing Fungi

The PUFA production from fungal can be realized by both submerged and solid-state fermentation. However, the metabolic fate of the substrate is primarily determined by the type of fermentation employed (Costa et al. 2018). Industrial production of PUFAs is largely carried out in submerged mode due to the ease of scale-up, recovery of biomass and high production yield with less fermentation time (Mamani et al. 2019). However, the large amount of wastewater generated and high energy requirement for the process have made the process costly and unsustainable (Asadi et al. 2015). Aerobic fermentation of a solid substrate by microorganisms in the absence/very low amount of free water to produce certain value-added bioproducts is referred to as solid-state fermentation (Dulf et al. 2020). Solid-state fermentation (SSF) application in food is an age-old practice.

Cheese, koji and tempeh are all traditional solid-state fermented food products (Rosales et al. 2018). However, in recent years solid-state fermentation by fungi has received growing interest among researchers especially in the food industry for enrichment of solid food with desired nutrients. Japan is one of the leading countries that has exploited SSF in the world with large-scale production plants (Gowthaman et al. 2001). Filamentous fungi are primarily exploited in SSF due to their ability to tolerate low water condition. In food industries, the ability of filamentous fungi to inhibit anti-nutrients and hydrolyze a number of substrates expands its potential to produce valuable food products from previously underutilized substrates. A number of agro-industrial wastes such as whey, wheat bran, grape pomace and apple byproducts have shown promising results for production of PUFA in SSF (Dulf et al. 2020). The PUFA-rich oil produced in this manner can find application as emulsifiers, as blends with other oils or as direct food supplements. Besides, the already edible solid food can also be enriched with PUFA using GRAS fungi and consumed directly (Diwan and Gupta 2019). FAOSTAT 2019 data indicate that worldwide cereals provide 50% of dietary energy in humans (Klempová et al. 2020). However, cereals have low lipid content but a good C/N ratio ranging from 20 to 60 (Čertík et al. 2013). Therefore, their nutritional value can be improved by fortifying them with functional lipids. To improve PUFA production of solid substrates in SSF, mainly three different strategies are employed:

- Gradual increase in C/N ratio by addition of a carbon source such as glucose/whey
- Optimization of physical conditions such as temperature, oxygen availability and water activity
- Addition of exogenous oil (Sláviková and Čertik 2005)

Nitrogen is essential for fungi growth especially during stationary phase. Organic nitrogen is a more preferred source for fungi in contrast to inorganic nitrogen (Asadi et al. 2015). However, after a certain period, high C/N ratio favours increased PUFA formation. GLA formation is highly elevated at a C/N ratio of around 160 (Dulf et al. 2020). Therefore, gradual addition of different carbon sources is preferred at different growth stages of fungi. It was reported that glucose supplementation of 30 and 40% in SSF of cereals by *M. circinelloides* CCF-2617 increased GLA yields as high as 3.5 times in comparison to control without glucose addition (Čertík et al. 2013). In a column reactor, high temperature and low oxygen availability at the bottom of the chamber are often the most encountered problems. A temperature in the range of 23–30 °C often promotes fungal growth. However, lower temperatures can substantially increase PUFA production (Asadi et al. 2015). Therefore, proper aeration and temperature maintenance are important parameters that need to be regulated for high product yield. Spent malt grain (SMG) is often used in SSF in a substrate: SMG ratio of 3:1. It has been found to act as an inert material highly improving the oxygen availability in the medium. Addition of SMG has also been shown to positively correlate with increased PUFA production (Čertík et al. 2013). Moisture content regulation is also another important aspect of SSF. Low moisture can result in substrate swelling along with decreased levels of nutrient solubility, while high moisture causes stickiness in substrate particles, thereby resulting in decreased levels of gaseous exchange and porosity. The preferred moisture content of substrate in SSF is 60–65 and 70–75% for ω-3 and ω-6 fatty acid synthesis, respectively. A pH of not more than 8 and less than 5, i.e. pH in the range of 6–7, is usually preferred for PUFA production (Asadi et al. 2015). It is also possible to control the direction of higher fatty acid synthesis by supplementing the substrate with oils. The addition of exogenous oils such as linseed oil and sunflower oil can direct lipid synthesis in favour of omega-3 (EPA) fatty acid synthesis, while sesame seeds promote omega-6 fatty acid synthetic pathway. Both linseed and sunflower oils, direct precursors of omega-3 fatty acid pathway, contain alpha-linoleic acid. Sesame seeds are known to contain inhibitors of delta-5 desaturase enzyme. Nutritionists and dieticians often recommended ω-6/ω-3 ratio of PUFA to be 5:1 or less (Patel et al. 2020). Therefore, this strategy can also be applied to balance ω-6/ω-3 ratio of food products as in *Mortierella alpina* which normally produces arachidonic acid and can be also made to produce EPA by addition of linseed oil (Asadi et al. 2015; Sláviková and Čertik 2005).

7.5   Processing Methods for Extraction of PUFA-Rich Oil

The PUFA-rich oil produced by fungal fermentation often undergoes downstream processing steps of lipid recovery and refining for application in food and other nutraceutical products. Two methods of oil recovery are currently employed for lipid recovery: dry methods that use solvents for extraction of lipids from dried cell mass and wet methods where the biomass is directly treated with enzymes without the use of solvents. Dry methods mainly include initial physical separation of biomass from culture media employing techniques such as centrifugation, flocculant addition and sedimentation depending upon the viscosity and composition of the culture media. The separated biomass is then dewatered by techniques such as spray drying, heating or freeze drying followed by the addition of an organic solvent. The mixture is then de-solventized to obtain a crude lipid extract and the solvent is generally recycled. To improve the efficiency of lipid extraction, other methods of fungal cell wall disruption such as acid treatment, microwave-assisted solvent extraction and physical disruption of fungal cell wall are also frequently employed in dry methods. A general outline of the above process is given in Fig. 7.3. The lipids produced by these methods are not pure and more prone to oxidation. Therefore, further refining of these oils is carried out by techniques such as degumming, bleaching and alkali refining (Khot et al. 2020; Ji et al. 2015; Ji and Ledesma-Amaro 2020; Cheng et al. 2019).

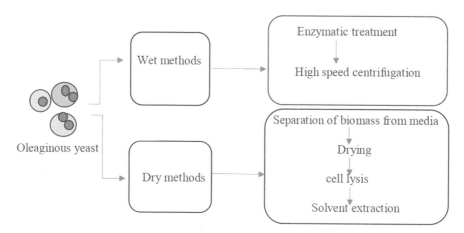

Fig. 7.3 General outline of a typical fungal lipid recovery process

7.6   Application in Food Industries

Linoleic acid (LA) and alpha-linoleic acid (ALA) are the precursors of ω-6 and ω-3 chain of PUFA, respectively. However, mammals lack the enzyme delta-12 desaturase that converts oleic acid to LA. Thus, both these fatty acids LA and ALA are regarded as essential in our diet from which other PUFAs can be derived in body. However, it has been found that this conversion efficiency in humans is very low. In females, only 10% of ALA is converted to EPA or DHA. Since human milk is a rich source of a number of PUFAs for a growing infant, sufficient intake of lipids is recommended for lactating mothers (Ganesan et al. 2014). Besides it was revealed that the blood levels of ARA in breast-fed infants are comparatively higher than infants fed with formula milk (Stahmann 2011). Since 90% of brain essential fatty acids are ARA and DHA, food formulations especially infant formulas are often supplemented with ARA-rich oil for better development of brain and nerve. The current market trends project that by 2025 industrial ARA production demand can reach an estimated 410 thousand tons. *Mortierella alpine* has been commercially exploited as a good source of ARA. ARASCO, RAO and SUNGTA are some commercially available oils with more than 40% ARA content (Mamani et al. 2019). The FDA of the USA has also given GRAS status to an oil combination of two-volume ARA and one-volume DHA for providing PUFA to infants (Ratledge 2013).

Ω-3 fatty acids such as EPA and DHA have also been implicated in a number of health-related benefits such as weight, cardiovascular and Alzheimer disease management (Swanson et al. 2012). An EPA oil rich in supplement for humans has been developed by DuPont. It is produced by genetically engineered *Yarrowia lipolytica* cells and sold under the trade name New Harvest™. The product contains purified lipids from the biomass. The EPA-rich biomass is also used to raise farmed salmon (Xie et al. 2015). Farmed fishes usually have lower levels of ω-3 PUFA due to their cultivation on vegetable oils that lack essential PUFAs that are available in wild. However, fishes raised with PUFA-fortified feeds can have a fatty acid composition similar to their wild counterparts which can ultimately raise the quality of meat obtained from them (Kwasek et al. 2020).

Targeting the already established system of food production with modern metabolomic engineering strategies can provide an effective mean of fortifying essential lipids in our diet. Fluxome Sciences A/S, a Denmark-based company, has developed a PUFA-producing metabolically engineered strain of *Saccharomyces cerevisiae* and used it in place of baker's yeast strain that is commercially available. The bread and fermented drinks prepared using this strain showed an improved PUFA profile in comparison to control. Besides, these single cells can also be consumed as whole (Plate et al. 2009).

7.7   Conclusion

The presence of lipid-synthesizing apparatus in fungi coupled with the fact that some of these microbes can even accumulate 60% of their cell dry weight as lipids is what makes them an attractive candidate for PUFA synthesis. Still the commercial production of PUFA from fungi is not economical. The main economic burden is the carbon source which is further poorly utilized by current available fungal strains. Therefore, to realize the true potential of fungi for PUFA production, it is essential to have a deeper understanding of the underlying lipid regulatory mechanisms operative in these microorganisms. This understanding can give useful insights for the approaches needed to direct the lipid-synthesizing machinery of these fungi to produce PUFA more efficiently. In the current scenario, a range of biological tools are available at hand to manipulate the genome of an organism and thus alter its metabolic profile. These tools can be utilized to develop strains that have greater product yield, have the ability to utilize carbon from waste and are non-toxic. An inter-disciplinary approach applying knowledge from all domains including system biology, synthetic biology and metabolic engineering is therefore the current need of the hour in microbial lipid research.

References

Akpinar-Bayizit A (2014) Fungal lipids: the biochemistry of lipid accumulation. Int J Chem Eng Appl 5(5):409–414

Asadi SZ, Khosravi-Darani K, Nikoopour H, Bakhoda H (2015) Evaluation of the effect of process variables on the fatty acid profile of single cell oil produced by *Mortierella* using solid-state fermentation. Crit Rev Biotechnol 35(1):94–102

Bellou S, Triantaphyllidou IE, Aggeli D, Elazzazy AM, Baeshen MN, Aggelis G (2016) Microbial oils as food additives: recent approaches for improving microbial oil production and its polyunsaturated fatty acid content. Curr Opin Biotechnol 37:24–35

Čertík M, Adamechová Z, Guothová L (2013) Simultaneous enrichment of cereals with polyunsaturated fatty acids and pigments by fungal solid state fermentations. J Biotechnol 168(2):130–134

Cheng MH, Rosentrater KA, Sekhon J, Wang T, Jung S, Johnson LA (2019) Economic feasibility of soybean oil production by enzyme-assisted aqueous extraction processing. Food Bioprocess Technol 12:539–550

Costa JAV, Helen H, Kumar V, Pandey A (2018) Advances in solid-state fermentation. In: Pandey A, Larroche C, Soccol CR (eds) Current developments in biotechnology and bioengineering: current advances in solid-state fermentation. Elsevier, pp 1–17

Czumaj A, Śledziński T (2020) Biological role of unsaturated fatty acid desaturases in health and disease. Nutrients 12:356

Diwan B, Gupta P (2019) Lignocellulosic biomass to fungal oils: a radical bioconversion toward establishing a prospective resource. In: Yadav A, Singh S, Mishra S, Gupta A (eds) Recent advancement in white biotechnology through fungi. Fungal biology. Springer, Cham, pp 407–440

Dulf FV, Vodnar DC, Toșa MI, Dulf EH (2020) Simultaneous enrichment of grape pomace with γ-linolenic acid and carotenoids by solid-state fermentation with Zygomycetes fungi and antioxidant potential of the bioprocessed substrates. Food Chem 310:125927

Ganesan B, Brothersen C, McMahon DJ (2014) Fortification of foods with omega-3 polyunsaturated fatty acids. J Crit Rev Food Sci Nutr 54(1):98–114

Gemperlein K, Dietrich D, Kohlstedt M, Zipf G, Bernauer HS, Wittmann C, Wenzel SC, Müller R (2019) Polyunsaturated fatty acid production by *Yarrowia lipolytica* employing designed myxobacterial PUFA synthases. Nat Commun 10:4055

Gill I, Valivety R (1997) Polyunsaturated fatty acids, Part 1: Occurrence, biological activities and applications. Trends Biotechnol 15(10):401–409

Gowthaman MK, Krishna C, Moo-Young M (2001) Fungal solid state fermentation – an overview. Appl Mycol Biotechnol 1:305–352

Ji X-J, Huang H (2019) Engineering microbes to produce polyunsaturated fatty acids. Trends Biotechnol 37(4):344–346

Ji X-J, Ledesma-Amaro R (2020) Microbial lipid biotechnology to produce polyunsaturated fatty acids. Trends Biotechnol 38(8):832–834

Ji XJ, Ren LJ, Huang H (2015) Omega-3 biotechnology: a green and sustainable process for omega-3 fatty acids production. Front Bioeng Biotechnol 3:158

Khan MAK, Yang J, Hussain SA, Zhang H, Liang L, Garre V, Song Y (2019) Construction of DGLA producing cell factory by genetic modification of *Mucor circinelloides*. Microb Cell Factories 18(1):64

Khot M, Raut G, Ghosh D, Alarcón-Vivero M, Contreras D, Ravikumar A (2020) Lipid recovery from oleaginous yeasts: perspectives and challenges for industrial applications. Fuel 259:116292

Kim SH, Park JS, Kim SY, Kim JB, Roh KH, Kim HU, Lee KR, Kim JB (2014) Functional characterization of polyunsaturated fatty acid delta 6-desaturase and elongase genes from the black seabream (*Acanthopagrus schlegelii*). Cell Biochem Biophys 68(2):335–346

Klempová T, Slaný O, Šišmiš M, Marcinčák S, Čertík M (2020) Dual production of polyunsaturated fatty acids and beta-carotene with *Mucor wosnessenskii* by the process of solid-state fermentation using agro-industrial waste. J Biotechnol 311:1–11

Kosa G, Vuoristo KS, Horn SJ, Zimmermann B, Afseth NK, Kohler A, Shapaval V (2018) Assessment of the scalability of a microtiter plate system for screening of oleaginous microorganisms. Appl Microbiol Biotechnol 102:4915–4925

Kothri M, Mavrommati M, Elazzazy AM, Baeshen MN, Moussa TAA, Aggelis G (2020) Microbial sources of polyunsaturated fatty acids (PUFAs) and the prospect of organic residues and wastes as growth media for PUFA-producing microorganisms. FEMS Microbiol Lett 367(5):fnaa028

Kwasek K, Thorne-Lyman AL, Phillips M (2020) Can human nutrition be improved through better fish feeding practices? A review paper. J Crit Rev Food Sci Nutr. https://doi.org/10.1080/10408398.2019.1708698

Laoteng K, Čertík M, Cheevadhanark S (2011) Mechanisms controlling lipid accumulation and polyunsaturated fatty acid synthesis in oleaginous fungi. Chem Pap 65(2):97–103

Mamani LDG, Magalhães AIM Jr, Ruanb Z, Carvalhoa JCD, Soccola CR (2019) Industrial production, patent landscape, and market trends of arachidonic acid-rich oil of *Mortierella alpine*. Biotechnol Res Innov 3(1):103–119

McNeil BA, Stuart DT (2018) *Lipomyces starkeyi*: an emerging cell factory for production of lipids, oleochemicals and biotechnology applications. World J Microbiol Biotechnol 34(10):147

Ochsenreither K, Glück C, Stressler T, Fischer L, Syldatk C (2016) Production strategies and applications of microbial single cell oils. Front Microbiol 7:1539

Park YK, Nicaud JM, Ledesma-Amaro R (2018) The engineering potential of *Rhodosporidium toruloides* as a workhorse for biotechnological applications. Trends Biotechnol 36(3):304–317

Patel A, Karageorgou D, Rova E, Katapodis P, Rova U, Christakopoulos P, Matsakas L (2020) An overview of potential oleaginous microorganisms and their role in biodiesel and omega-3 fatty acid-based industries. Microorganisms 8(3):434

Plate I, Gunnarsson NK, Møller P, Geogieva TI, Stenhuus B, Inventor (2009) Fluxome Sciences A/S, assignee. Foods and beverages with increased polyunsaturated fatty acid content. World Patent WO 2009080630

Ratledge C (2013) Microbial oils: an introductory overview of current status and future prospects. OCL 20(6):D602

Rosales E, Pazos M, Sanromán MA (2018) Solid-state fermentation for food applications. In: Pandey A, Larroche C, Soccol CR (eds) Current developments in biotechnology and bioengineering: current advances in solid-state fermentation. Elsevier, pp 319–355

Sakuradani E, Ando A, Ogawa J, Shimizu S (2009) Improved production of various polyunsaturated fatty acids through filamentous fungus *Mortierella alpina* breeding. Appl Microbiol Biotechnol 84(1):1–10

Sláviková L, Čertik M (2005) Microbial preparation of polyunsaturated fatty acids by fungal solid state fermentation. Chem List 99:s234–s237

Smith W, Mukhopadhyay R (2012) Essential fatty acids: the work of George and Mildred Burr. J Biol Chem 287(42):35439–35441

Spector AA, Kim HY (2015) Discovery of essential fatty acids. J Lipid Res 56(1):11–21

Srianta I, Nugerahani I, Kusumawati N (2010) Production of polyunsaturated fatty acids with *Rhizomucor miehei* by submerged fermentation. Asian J Food Agro-Ind 3(2):293–300

Stahmann KP (2011) Production of Vitamin B2 and a polyunsaturated fatty acid by fungi. In: Hofrichter M (ed) Industrial applications. The Mycota (A comprehensive treatise on fungi as experimental systems for basic and applied research). Springer, Berlin/Heidelberg, pp 235–247

Swanson D, Block R, Mousa SA (2012) Omega-3 fatty acids EPA and DHA: health benefits throughout life. Adv Nutr 3(1):1–7

Verma ML, Kishor K, Sharma D, Kumar S, Sharma KD (2020) Microbial production of omega-3 polyunsaturated fatty acids. In: Verma ML, Chandel AK (eds) Biotechnological production of bioactive compounds. Elsevier, pp 293–326

Wang H, Zhang C, Chen H, Gu Z, Zhao J, Zhang H, Chen YQ, Chen W (2020) Tetrahydrobiopterin plays a functionally significant role in lipogenesis in the oleaginous fungus *Mortierella alpine*. Front Microbiol 11:250

Xie D, Jackson EN, Zhu Q (2015) Sustainable source of omega-3 eicosapentaenoic acid from metabolically engineered *Yarrowia lipolytica*: from fundamental research to commercial production. Appl Microbiol Biotechnol 99:1599–1610

Chapter 8
Fungal Production of Food Supplements

Archana Singh, Modhurima Misra, Shashank Mishra,
and Shashwati Ghosh Sachan

8.1 Introduction

In the United States, according to the Dietary Supplement Health and Education Act (DSHEA) of 1994, a dietary supplement is a product intended to supplement (reinforce) the diet which contains a 'dietary ingredient', such as minerals, vitamins, amino acids, enzymes, organ tissues, glandulars, herbs or other botanics, or a concentrate, metabolite, constituent, extract or combination of any of the above (DSHEA 1994; Questions and answers on dietary supplements n.d.). Dietary supplements can be in the form of tablets, capsules, lozenges, chewables, powders, solutions or syrups. They may also appear as part of foods and can be categorized into natural supplements that are extracted from plants, animal tissues or inorganic material, such as seawater and rocks; semi-synthetic supplements extracted from natural sources and then chemically changed; and synthetic supplements that are completely artificially produced. Practically every nutrient can appear as a dietary supplement which mainly involves minerals, vitamins, fibres, prebiotics, amino acids, proteins, phytonutrients (polyphenols and others), omega-3 fatty acids and conjugated linoleic acids.

A rapid switch towards the bio-production and use of compounds of (micro) biological origin is observed due to increasing consumer concern and interest

---

Shashank Mishra: Co-corresponding author.

A. Singh · M. Misra · S. G. Sachan (✉)
Department of Bio-Engineering, Birla Institute of Technology, Mesra,
Ranchi, Jharkhand, India
e-mail: ssachan@bitmesra.ac.in

S. Mishra (✉)
Quality Control and Quality Assurance Laboratory, Biotech Park,
Lucknow, Uttar Pradesh, India

© The Author(s), under exclusive license to Springer Nature
Switzerland AG 2021
X. Dai et al. (eds.), *Fungi in Sustainable Food Production*, Fungal Biology,
https://doi.org/10.1007/978-3-030-64406-2_8

towards the use of natural and environment-friendly products especially when related to food and home-care products (Cheetham 1997). According to US as well as European regulations, products, obtained through microbial or enzymatic processes, are considered as natural. Human affairs have been influenced by fungi for thousands of years, whether as a direct food source, as a medicine, or as a processed food form, due to its nutritive value (Moore et al. 2011). The presence of essential and non-essential amino acids in the fungi has a profound influence on the nutritional value of fungi as food and feed. Today food and food supplements as well as other dietary sources and fermented beverages obtained from the fungal origin are consumed all over the world in vast quantities which leads to its commercial production by a rapidly growing industry. However, the consumption of wild fungi was first reliably noted in China, several hundred years before the birth of Christ (Aaronson 2000).

In spite of having various evidences and age-old practice of consumption of fungi and their utilization in production of other food materials, their full potential was not explored until the latter half of the twentieth century when its potential application as dietary source was boosted by the advent of the golden age of industrial microbiology (Aaronson 2000). Since then, this diverse community represented by yeasts, mushrooms and filamentous fungi has been exploited in numerous food products being used for both human and livestock consumption. In the beginning of the seventeenth century, the cultivation of macrofungi to yield fruit bodies began to flourish, and today over 6 million tonnes of edible mushrooms are produced commercially each year around the world. However, all of the wild varieties of macrofungi are not propagated on a commercial scale. The choice of the fungal strain used in the food industry is based on their production yields and regulatory issues, especially fungal strains which have attained the so-called GRAS (Generally Recognized As Safe) status, by the US Food and Drug Administration (FDA) are preferred among others.

Besides being used as dietary sources and other food and feed supplements, various other fungi have been reported to modify food to make it more nutritious or palatable. For example, soy sauce being used as flavouring and condiment in Chinese cuisine was produced by growing filamentous fungi such as *Aspergillus oryzae* and *A. sojae* on cooked soya beans. Similarly, *Tuber melanosporum* is used as flavouring instead of a separate dish known to most as the truffle. There are certain natural advantages of the fungal kingdom in terms of their dietary supremacy over the rest of the vegetarian platter. These are: (a) a good protein content (20–30% of dry matter) having all the essential amino acids (yeasts are especially enriched in lysine) and thus capable of substituting meat, (b) having chitinous wall to act as a source of dietary fibre, (c) high vitamin B content, (d) low in fat, (e) virtually free of cholesterol and (f) low production cost (Breene 1990).

Use of agro-wastes and other industrial waste products as substrate leads to the low production cost for mushroom cultivation, for example, use of cotton wastes for the cultivation of oyster mushroom species (*Pleurotus ostreatus*, *Pleurotus cystidiosus*, *Pleurotus sajor-caju*) (Cohen et al. 2002). Similarly cotton wastes can also be used for the production of the straw mushroom (*Volvariella volvacea*) which was traditionally grown in Southeast Asia as rice straw (Hsu et al. 1997). On the other hand, all other agricultural production generates enormous waste since little of each crop is actually used such as 5% in palm and coconut plantation, 2% in sisal plantation and 7% in sugarcane plantation. Many lignocellulose agricultural wastes for the growth of *Pleurotus* sp. also help in bioremediation and production of a consequent cash crop in the form of mushroom (Ghorai et al. 2009).

8.2 Use of Fungi in Dietary Food

Over recent years, due to increased public concern about dietary and health issues, consumption of fungal food in the form of either freshly cooked mushrooms or processed foods, beverages and dietary supplements of fungal origin has increased on a global basis.

8.2.1 Use of Fruiting Body

Fresh or processed fruiting bodies of mushrooms have been consumed directly and used as delicacy (Moore and Chiu 2001). Fermentative processes through stages of media preparation, inoculation and incubation were used for the production of fungi. Substrates available from cheap valued sources like agro-biomass and industrial wastes were used in the production media and transformed into high value-added food and pharmaceutical products. Thus, use of fungi is important from economical as well as environmental aspects. Various species of edible mushrooms exist in the wild, out of which less than 20 species are used extensively as food and only 8–10 species are regularly cultivated to any significant extent. The most commonly eaten species, *Agaricus bisporus*, sold as button mushrooms when small or portobello mushrooms when larger, are used in salads, soups, and many other dishes (Beelman et al. 2003). The steps of commercial mushroom production in the European tradition involve composting, spawning, casing and cropping (Moore and Chiu 2001).

Many commercially grown Asian fungi are now available and have gained huge popularity in the West. Various mushroom varieties such as straw mushrooms (*Volvariella volvacea*), oyster mushrooms (*Pleurotus ostreatus*), shiitakes (*Lentinula edodes*) and enokitake (*Flammulina* spp.) can be freshly availed from grocery stores and markets. They are often used for the preparation of various types of dishes. There are many other fungi like milk mushrooms (*Lactarius deliciosus*), morels (*Morchella esculenta*), chanterelles (*Cantharellus cibarius*), truffles, black trumpets (*Craterellus cornucopioides*) and porcini mushrooms *(Boletus edulis)* (also known as 'king boletes') available in the market as a vegetarian substitute of protein source (Murata et al. 2002; Rossi et al. 1993; She et al. 1998; Cheung 1996; Cheung et al. 2003; Hsu et al. 1997; Ko et al. 1995; Carbonero et al. 2008). All of them have a very high market price. The fruiting body of fungi can be cultivated through horticulture.

After *Agaricus*, 'shiitake mushroom' (*Lentinula edodes*) is the second most cultivated mushroom in the world. Besides China and Japan, shiitake is also widely cultivated in Taiwan, Thailand, Korea, Singapore as well as Holland, the United States and Canada. The 'shiitake mushroom' is as common in Asian countries as *Agaricus bisporus* as in the West. Shiitake's protein can be used extensively in a vegetarian diet due to a full complement of essential amino acids. Its active ingredient, lentinan (a polysaccharide), was found to show activity against cancer and cholesterol (Beelman et al. 2003). The most commonly used cultivated edible macrofungi and their nutritional properties are listed in Table 8.1.

8.3 Use of Fungi as and in Processed Food

Various food and food additives in the markets being used as animal feed or human food are constituted by fungi.

8.3.1 SCP

A variety of microbial products, produced by fermentation that can be used for the fermentation of vast amounts of waste materials, such as straws; wood and wood-processing wastes; food, cannery and food-processing wastes; and residues from alcohol production or from human and animal excreta, have been given the name single cell protein. Single cell protein (SCP) is considered as the most promising nonconventional protein source that can be used as the dietary component for augmenting the anticipated world protein shortage (Moore et al. 2001). Yeast is an especially important fungus and considered as the most promising source of SCP although its growth is not as rapid as that of bacteria. Production of SCP involves the use of various waste products, including cheese whey, starch, fruit-processing residues, animal waste and petroleum hydrocarbons. The study carried out by Lyutskanov et al. (1990), on the applicability of protein extracts obtained from a strain of *Saccharomyces cerevisiae*, revealed that the percentage of essential amino acids lysine and isoleucine was higher than that of soya bean and the chemical score of these amino acids is higher than that of egg protein (Lyutskanov et al. 1990). The preparation of the protein extract using yeast is associated with the problem of high nucleic acid content which can cause gout and urinary tract stones which can be reduced using successful methods of reduction (Moore and Chiu 2001). It was also found that the use of nonconventional protein sources with high carbohydrate content shows allergic reactions. Low nucleic acid and carbohydrate content of yeast

Table 8.1 Nutritional properties of commonly used edible macrofungi

Fungi (Common name)	Appearance	Cultivation method	Nutritional properties	References
Lentinula edodes (shiitake or shiang-gu)	Dark brown cap with white stalk	Shiitakes grow in groups on the decaying wood of deciduous trees. The principal method of cultivation involves inoculation of wooden logs with spore inoculum or mycelial plugs, then the logs are allowed to stand for up to 9 months to achieve colonization by the fungus	Contains carbohydrate, protein, low in fat, high percentage of polyunsaturated fatty acids; also contains many vitamins, adenine and choline content effective in preventing the occurrence of cirrhosis of the liver as well as vascular sclerosis	Murata et al. (2002) and Rossi et al. (1993)
Volvariella volvacea(straw mushroom or paddy straw mushroom)	Pink gills and spore prints; lack a ring; have an *Amanita*-like volva at the stem base. The gills of young *Volvariella* are white	Grown on rice straw beds and picked immature, during the button or egg phase and before the veil ruptures, take 4 to 5 days to mature	Natural source of antioxidant due to high β-carotene content; contains protein, fat, iron, zinc, various amino acids and large amounts of vitamin C	She et al. (1998), Cheung (1996) and Cheung et al. (2003)
Flammulina velutipes (winter mushroom)	Convex-shaped cap; moist and sticky when fresh; colour variable –dark orange brown to yellowish brown; gills attached to the stem and whitish to pale yellow, becomes dark rusty brown on maturity	Grown in dark or cold places; stems are forced to stretch by placing a tightly fitting collar around clusters of mushroom	Mannofucogalactan, a heterogalactan derived from *Flammulina*, is known to possess nutritional values	Hsu et al. (1997), Ko et al. (1995) and Carbonero et al. (2008)

(continued)

Table 8.1 (continued)

Fungi (Common name)	Appearance	Cultivation method	Nutritional properties	References
Agaricus bisporus (button mushroom, champignon)	The original wild form bears a brownish cap and dark brown gills but the more familiar ones are with a white cap, stalk and flesh and brown gills	Substrate for growth is straw; composted with animal manures and other organic nitrogen compounds over a period of 1 to 2 weeks; the final product is a unique substrate suitable for the rapid growth of the *Agaricus* inoculum	Fairly rich in vitamins like vitamin B and minerals like sodium, potassium, phosphorus and selenium; rich in protocatechuic acid and pyrocatechol. Raw mushrooms are naturally cholesterol- and fat-free	Beelman et al. (2003) and Rossi et al. (1993)
Tuber melanosporum (truffle)	Fruiting body or truffle is round, pitted and white when young but darkens as it matures	Cultivation involves the planting of hazel trees whose roots are inoculated with truffle mycelium. The first fruiting bodies can be harvested about 4 to 10 years after planting the trees	Truffle has tantalizing taste and aroma and is most sought after delicacy with great economic value. Truffle oil is used as a lower-cost and convenient substitute for truffles, to provide flavouring, or to enhance the flavour and aroma of truffles in cooking	Breene (1990)
Ganoderma lucidum (reishi)	Large, hard and leathery fungus with sessile or stalked basidiocarps having tiny pores undersurface	Cultivated on hardwood logs or sawdust-/woodchip-based formulations; attempt was made to use paddy straw as a substrate to cultivate *G. lucidum*	Used in dietary preparation and to make tea or soup. Protein comprises only 7.3% of dry weight. Glucose accounted for 11% and metals 10.2% of dry mass (K, Mg, Ge and Ca being the major trace components), produces a group of triterpenes called ganoderic acids, which have a molecular structure similar to steroid hormones	Bao et al. (2002), Wang et al. (2006) and Zhang et al. (2002)
Morchella elata (the black morel)	Grey ridges or tan; pits are brown and elongated when young; turn black with age	Exhibit a pyrophilic behaviour and may grow abundantly in forests which have been recently burned by a fire	Excellent source of vitamin D2	Mattila et al. (2000)

make it a good candidate for large-scale production of high nutritional quality protein extracts, potentially applicable to human nutrition. One example could be to increase the protein quality of cereal-based food products. Generally, SCP has been obtained in a diluted form after fermentation, containing less than 5% solids, which is further concentrated (Fig. 8.1).

The major uses of yeasts have been highlighted as follows.

8.3.2 Baker's Yeast

The production of baker's yeast for food purposes involved the largest domestic use of a microorganism. Baker's yeast is a strain of *Saccharomyces cerevisiae* which is mixed with bread dough to bring about vigorous sugar fermentation (Taniwaki et al. 2001). The carbon dioxide produced during the fermentation leads to leavening or rising of the dough. The selection of the strain is based on its capacity to produce abundant gas quickly, its viability during ordinary storage and its ability to produce desirable flavour.

8.3.2.1 Use of Yeast Cells in Food and Fodder

The definition of food yeast was proposed by Professor Jacquot and Dr. Biloraud in 1957 which was subsequently adopted by the I.U.P.A.C. and then by the European Economic Community in 1975. It has been defined as 'yeast that has been killed and dried is known as food yeast and it should have no diastase activity and should be free from food additive'. Fodder yeasts are the yeasts acting as food supplements of domestic livestock. Essentially, these organisms must comply with certain nutritional properties of vitamin content, protein content, amino acid composition, good digestibility and absence of toxic substances intended for human and animal consumption (Ghorai et al. 2009).

Fig. 8.1 Dehydration methods used for the processing of SCP

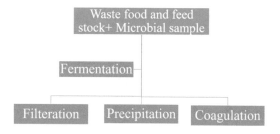

8.4 Processed Fungal Food as an Alternative to SCPs

8.4.1 Mycoprotein

Various popular processed food products are available in the market. The most notable one is the mycoprotein 'Quorn'. The UK Foods Standards Committee coined the term mycoprotein to serve as the generic name for a food product obtained using continuous fermentation of a selected strain of *Fusarium venenatum* (originally called *F. graminearum*). Very large air-lift fermenter has been used to grow the mycelium in a continuous culture mode (Trinci 1992). During continuous fermentation, mycoprotein is grown on food-grade glucose derived from starch as the carbon source. A filamentous hyphal mass is the resulting product of the above reaction much resembling a meat-like texture which was originally dried and powdered for sale as high-protein SCP flour, but due to organoleptic qualities of the hyphal mass, it has been developed as a meat substitute under the brand name 'Quorn', a high-technology product coupled with the inherent nutritional value of fungal biomass. Complementary to beef, the product is a low-fat, low-calorie, cholesterol-free health food (Peinter et al. 1998).

The nutritional value of both mycoprotein and mushroom is comparable. It is rich in protein and dietary fibre (due to the presence of chitin and β-glucans in hyphal walls) and contains adequate proportions of all essential amino acids. It has a protein content minimum of 44%. Its net protein utilization (NPU) is 75/100, comparable to beef (80) and cow's milk (75). When supplemented with 0.2% methionine, NPU is 100 (equal to egg protein). Mycoproteins are good source of B vitamins such as biotin (16 mg/100 g), riboflavin, niacin and B6 (content equivalent to lamb), although it lacks in B12. Mycoprotein is found to be the excellent source of zinc, particularly useful to vegetarians (Moore et al. 2001).

8.5 Production of Bakery and Cheese Products

Mixing of flour (usually from cereals especially wheat) with water, salt and sugar followed by leavening using yeast results in the production of bakery products. Fermentation of the sugar by the yeast forms carbon dioxide and alcohol. Bubbles are formed by the released gas due to the elastic extension of gluten (a protein) in the flour. On baking, the alcohol evaporates. The texture and flavour of the bread are determined by the length of leavening, the quantity of gluten in the flour, the constituents of the grain and the temperature (Taniwaki et al. 2001).

A solid or semisolid protein food product manufactured from milk is considered as cheese and is an occidental equivalent of the fermented soya products which are popular in Asia. Cheese manufacture was the only method of preserving milk before the advent of modern methods of food processing, like refrigeration, pasteurization and canning. Although bacterial fermentation is involved in basic cheese-making, filamentous fungi contribute in two important processes; these are the provision of enzymes for coagulation and mould-ripening (Freitas and Malcata 2000). The action of enzymes is involved in cheese production which coagulates the proteins in milk, forming solid curds (from which the cheese is made) and liquid whey, and animal enzymes, specifically chymosin and pepsin, extracted from the stomach membranes of unweaned ruminants were used in traditional cheese-making.

Very recently, the market has been dominated by the cheese products obtained using animal enzymes produced by genetically modified microbes, but for the coagulation step, the industrial cheese production still depends on enzymes from filamentous fungi (Gupta et al. 2003). Two of the most familiar examples of blue cheese are 'Camembert' and 'Roquefort'. Two species of *Penicillium* were used for their cheese production, *P. camemberti*, in Camembert cheese and *P. roqueforti* in Roquefort cheese, respectively. The cheese making process started approximately 4000 years ago involving the addition of moulds and other microorganisms for adding flavour to cheese. The processes are usually referred to as mould ripening. Methyl ketones, particularly 2-heptanone, are produced as the major flavour and odour compounds during the fungus growth throughout the cheese. Camembert and Brie are ripened by *Penicillium camemberti*, which produces changes in the texture of the cheese rather than its flavour. Extracellular proteases released by the fungus during its growth on the surface of the cheese help in the digestion of the cheese to a softer consistency from the outside towards the centre (Moore and Chiu 2001).

8.6 Use of Fungi in Beverage Production

Many alcoholic beverages are produced using the widely used yeast strain *Saccharomyces cerevisiae* which have been classified into three categories: those produced using fruit juices, those produced using starchy materials and those produced using other plant materials (Carlile et al. 2001). Production of alcoholic beverages using fruit juices includes wine, cider and perry. Wine is the product of the alcoholic fermentation by *S. cerevisiae* of the fresh grape juices and produced by fermenting red and white grapes, where the yeast converts the fruit sugars into alcohol. In Germany, *Botrytis cinerea*, a grape vine pathogen, is used to produce a particular wine 'Sauternes' or 'Edelfaule'. The pathogen attacks the growing grapes during their fermentation by growing their hyphae into the skin of the grapes causing water loss, resulting in shrivelled grapes with unusually high sugar concentration. The concentration of the sugars inside leads to the removal of any residual acidity, and the grapes are then harvested and fermented using the traditional methods, resulting into wine of sweet and high quality. Apples are used for the production of cider that can be classified into sweet (low in acid and tannin), bittersweet (low in acid, high in tannin), sharp (high acid, low tannin) and bitter sharp (high in both acid and tannin). Apple contains less sugar than grapes, due to which there is a decreasing alcohol concentration in cider. Perry pears or dessert pears are used for the production of perry. Same techniques and organism for fermentation are used for the production of cider from apple juices and perry from the juices of pears. The most common food and beverages obtained using fungi are listed in Table 8.2.

Beer is one of the widely used alcoholic beverages made from malted barley and flavoured with hops (female flowers of *Humulus lupulus*). In the beer industry, malting is used to achieve the degradation of starchy materials into simple sugars. Malting is done before fermentation by adding warm water to the coarse flour of malt for the initiation of barley enzymes to break the starch and proteins in the malt followed by boiling to stop enzymatic activities and the hop is mixed to add flavour. Beers are basically of two main types: ale and lager. In the traditional fermenter at room temperature, ale is produced by the top yeast followed by its mixing with gas accumulated as foam at the top of the vat during ale production and is restricted to British Isles. On the other hand, the bottom yeast is used for the production of lager which remains at the bottom because of their more hydrophilic cell walls which tend them to sediment and is served well-chilled throughout the world. The lager yeast is *S. carlsbergensis* and the ale yeast is *S. cerevisiae* in the brewing industry. After maturation, fining is done for the clarification of beer to get the final product (Ghorai et al. 2009). In Japan, rice is used as a starting material for the production of 'sake'. Rice is rich in starch and thus not utilizable by the yeast. Therefore, a fungus, *Aspergillus oryzae*, is allowed to grow on the surface of the rice. The fungus converts the starch into simpler sugars, which then can be used by the yeast for the production of sake. 'Sake' is inoculated with *Aspergillus oryzae* to produce 'koji'.

Various bacterial species were also used along with fungi for the production of non-alcoholic beverages like 'coco' (*Leuconostoc mesenteroides*, *Candida* sp.) and coffee (*Leuconostoc mesenteroides*, *Saccharomyces marscianus*, *Flavobacterium* spp., Fusarium spp.) (Soccol et al. 2008; Pasha and Reddy 2005).

8.7 Production of Other Food Products/Condiments/ Additives

Besides being used directly as food, fungi are also used in the processing of various food products. Research has been carried out on some of the fermented food products using the microorganisms (including bacteria). But majority of the microorganisms involved in the fermented foods (approximately 500 in numbers) are unknown. Some of the more familiar ones include 'miso', 'shoyu', 'tofu' and 'tempeh'. In these applications the fungus is primarily responsible for the production of some characteristic flavour, odour or texture and may or may not become part of the final edible product. Fungal enzymes of the host are released into the respective substrates by the concerned fungi and result in the production of fermented foods (Cowan 1996; Knoss et al. 1998).

Table 8.2 List of food and beverages produced using fungi and used directly as food

Product type	Raw material	Species	Process	Commercial product
Miscellaneous industrial products		*Saccharomyces cerevisiae*	Enzyme synthesis, invertase	Used in preparation of soft-centre candies, e.g. cordial cherries
	Starch	*Aspergillus niger*	Aerobic metabolism	Citric acid
		Aspergillus niger	Produce alpha-d-galactosidase enzyme	Enzyme suppresses methane production in humans
	Starch	*Aspergillus sp.*	Synthesis of amylase enzymes	Used for bread making and textile fibres
		Aspergillus sp.	Synthesis of pectinase enzymes	Clarification of fruit juices
	Corn starch solution	*Penicillium notatum*	Aerobic metabolism	Penicillin
			Synthesis of enzyme, glucose oxidase	Removal of oxygen from canned fruits, dried milk and other products
	Corn starch solution	*Fusarium moniliforme*	Aerobic synthesis	Plant hormone-gibberellin
Cheeses	Milk curd	*Penicillium roqueforti*	Production of blue pigment	Roquefort cheese
	Milk curd	*Penicillium candidum and P. camemberti*	Aerobic metabolism	Brie, Camembert and Limburger cheeses
Alcoholic beverages	Germinated grain (malt)	*Saccharomyces sp.*	Natural fermentation	Beer
	Fruit juice		Natural fermentation	Wine
	Rice		Natural fermentation	Sake
	Fruit juice		Fermentation and distillation	Brandy
	Grain mash		Fermentation and distillation	Whiskey
	Molasses		Fermentation and distillation	Rum
	Potatoes		Fermentation and distillation	Vodka
	Agave		Fermentation	Tequila

8.7.1 Shoyu

One of the most familiar Asian food products is 'soya sauce' (shoyu or soy sauce) made by mixed cooked soybeans and wheat flour (Archer 2000; Dunn-Coleman et al. 1991), which is then pressed into cakes and inoculated with *Aspergillus oryzae*. The mixture of salt and water produced the moulded cake known as 'koji' and is referred to as the 'moromi' which is then inoculated with a bacterium, *Pediococcus soyae*, and yeasts *Saccharomyces rouxii* and *Torulopsis* sp., to ferment the mixture for approximately 6 months to turn into soy sauce.

8.7.2 Miso

The Japanese word 'miso' is used for fermented soybean paste used as a flavouring agent not usually consumed by itself, but as a base for soup (Pariza and Johnson 2001; Feng et al. 2007). 'Miso' fermentation consists of washed, polished rice, which is steamed and inoculated with *Aspergillus oryzae* resulting in 'rice koji' followed by the digestion of the carbohydrates and proteins of the inoculated rice to sugars and amino acids by the fungus. The 'rice koji' is then inoculated by yeasts and bacteria and allowed to ferment.

8.7.3 Indonesian Tempeh

Fermentation of partially cooked soybean cotyledons with *Rhizopus oligosporus* leads to the production of a white cake known as Indonesian tempeh. This product has been used as a meat substitute and is on an increase of being widely sold into the vegetarian market. This was produced using the fungus which binds the soybean mass into a protein-rich cake (Feng et al. 2007).

There are a variety of such fermented products. One such product is 'angkak' which is a rice product popular in China and the Philippines, which is fermented using *Monascus* species. The characteristic pigments and ethanol which are used for red rice wine and food colouring are produced using *Monascus purpureus*. A mixture of red, yellow and purple polyketides is present in the pigments and about ten times more pigments are obtained from solid state fermentation than from submerged liquid fermentation (Moore et al. 2001).

8.8 Conclusion

Strong upsurge of fungal community in the field of food, feed and therapeutics leads to the development of research initiatives aimed for the development of industrial processes focusing on the production fungal-based food products, food supplements and various nutraceuticals and condiments. Fungal community also acts as the versatile tools in the vast field of medical research and acts as the model research organism. Over the past few decades, mushroom cultivation has been a promising option in both the developed and developing countries for increasing both rural income and earning of foreign currency. Nowadays, fungal community is also used to cater to more and more diversified human needs of treating human ailments, bioremediation and bio-fuel production. Human populations are also influenced by the fungi on a large scale because they are part of the nutrient cycle in ecosystems. Besides being used as the food supplements, food additives as well as nutraceuticals, they also have other ecosystem uses, such as pesticides and biocontrol agents.

References

Aaronson S (2000) Fungi. In: Kiple KF, Ornelas KC (eds) The Cambridge world history of food. Cambridge University Press, Cambridge, pp 313–336

Archer D (2000) Filamentous fungi as microbial cell factories for food use. Curr Opin Biotechnol 11:478–483

Bao XF, Wang XS, Dong Q, Fang JN, Li XY (2002) Structural features of immunologically active polysaccharides from *Ganoderma lucidum*. Phytochemistry 59:175–181

Beelman RB, Royse D, Chikthimmah N (2003) Bioactive components in Button Mushroom *Agaricus bisporus* (j.lge) imbach (Agaricomycetideae) of nutritional, medicinal, or biological importance (review). Int J Med Mushrooms 5:321–337

Breene WM (1990) Nutritional and medicinal value of specialty mushrooms. J Food Prot 53:883–894

Carbonero ER, Gorin PAJ, Iacomini M, Sassaki GL, Smiderle FR (2008) Characterization of a heterogalactan: some nutritional values of the edible mushroom *Flammulina velutipes*. Food Chem 108:329–333

Carlile MJ, Watkinson SC, Gooday GW (2001) The fungi. Academic Press, San Diego/London

Cheetham PSJ (1997) Combining the technical push and the business pull for natural flavours. Adv Biochem Eng Biotechnol 55:1–49

Cheung CKP (1996) The hypocholesterolemic effect of two edible mushrooms: *Auricularia auricula* (tree-ear) and *Tremella fuciformis* (white jelly-leaf) in hypercholesterolemic rat. Nutr Res 16:1721–1725

Cheung LM, Cheung PCK, Ooi VEC (2003) Antioxidant activity and total phenolics of edible mushroom extracts. Food Chem 81:249–255

Cohen R, Persky L, Hadar YPL (2002) Biotechnological applications and potential of wood-degrading mushrooms of the genus *Pleurotus*. Appl Microbiol Biotechnol 58:582–594

Cowan D (1996) Industrial enzyme technology. Trends Biotechnol 14:177–178

Dietary supplement health and education act of 1994, Pub. L. No. 103-417, 103RD Congress National Institutes of Health, Office of Dietary Supplements

Dunn-Coleman NS, Bloebaum P, Berka R, Bodie E, Robinson R, Armstrong G et al (1991) Commercial levels of chymosin production by *Aspergillus*. Biotechnol 9:976–981

Feng XM, Larsen TO, Schnurer J (2007) Production of volatile compounds by *Rhizopus oligosporus* during soybean and barley tempeh fermentation. Food Technol 19:173–181

Freitas C, Malcata FX (2000) Microbiology and biochemistry of cheeses with appelation d'Origine protegee and manufactured in the Iberian Peninsula from ovine and caprine milks. J Dairy Sci 83:584–602

Ghorai S, Banik SP, Verma D, Chowdhury S, Mukherjee S, Khowala S (2009) A review on fungal biotechnology in food and feed processing. Food Res Int 42:577–587

Gupta R, Rathi P, Bradoo S (2003) Lipase mediated upgradation of dietary fats and oils. Crit Rev Food Sci Nutr 43:635–644

Hsu CI, Hsu HC, Kao CL, Lin JY, Lin RH (1997) Fip-vvo, a new fungal immunomodulatory protein isolated from *Volvariella volvacea*. Biochem J 323:557–565

Knoss W, Reuter B, Alkorta I, Garbisu C, Llama MJ, Serra JL (1998) Industrial applications of pectic enzymes: A review. Process Biochem 33:21–28

Ko JL, Hsu CI, Lin RH, Kao CL, Lin JY (1995) A new fungal immunomodulatory protein, FIP-fve isolated from the edible mushroom, *Flammulina velutipes* and its complete amino acid sequence. Eur J Biochem 228:244–249

Lyutskanov N, Koleva L, Stateva L, Venkov P, Hadjiolov A (1990) Protein extracts for nutritional purposes from fragile strains of *Saccharomyces cerevisiae*: reduction of the nucleic acid content and applicability of the protein extracts. J Basic Microbiol 30:523–528

Mattila P, Piironen V, Suonpaa K (2000) Functional properties of edible mushrooms. Nutrition 16:694–696

Moore D, Chiu SW (2001) Fungal products as food. In: Pointing SB, Hyde KD (eds) Bio-exploitation of filamentous fungi. Fungal Diversity Press, Hong Kong, pp 223–251

Moore D, Robson GD, Trinci APJ (2001) 21st century guidebook to fungi. Cambridge University Press, Cambridge

Moore D, Robson GD, Trinci APJ (2011) 21st century guidebook to fungi. Cambridge University Press, Cambridge

Murata Y, Shimamura T, Tagami T, Takatsuki F, Hamuro J (2002) The skewing to Th1 induced by lentinan is directed through the distinctive cytokine production by macrophages with elevated intracellular glutathione content. Int Immunopharmacol 2:673–689

Pariza MW, Johnson EA (2001) Evaluating the safety of microbial enzyme preparations used in food processing: update for a new century. Regul Toxicol Pharmacol 33:173–186

Pasha C, Reddy G (2005) Nutritional and medicinal improvement of black tea by yeast fermentation. Food Chem 890:449–453

Peinter U, Poder R, Pumpel T (1998) The iceman's fungi. Mycol Res 102:1153–1162

Questions and Answers on Dietary Supplements (n.d.) U.S. Food and Drug Administration. Available from: https://www.fda.gov/Food/DietarySupplements/UsingDietarySupplements/ucm480069.htm

Rossi V, Jovicevic L, Nistico V, Orticelli G, Troiani MP, Marini S (1993) In vitro antitumor activity of *Lentinus edodes*. Pharmacol Res 27:109–110

She QB, Ng TB, Liu WK (1998) A novel lectin with potent immunomodulatory activity isolated from both fruiting bodies and cultured mycelia of the edible mushroom *Volvariella volvacea*. Biochem Biophys Res Commun 247:106–111

Soccol CR, Dalla Santa HS, Rubel R, Vitola FM, Leifa F, Pandey A et al (2008) Mushrooms-A promising source to produce nutraceuticals and pharmaceutical byproducts. In: Koutinas AA, Pandey A, Larroche C, Larroche A (eds) Current topics on bioprocesses in food industry. Asiatech Publishers, Inc, New Delhi, pp 439–448

Taniwaki MH, Silva N, Banhe AA, Iamanaka BT (2001) Comparison of culture media, simplate, and petrifilm for enumeration of yeasts and molds in food. J Food Prot 64:1592–1596

Trinci APJ (1992) Myco-protein: a twenty-year overnight success story. Mycol Res 96:1–13

Wang XJ, Bai JG, Liang YX (2006) Optimization of multienzyme production by two mixed strains in solid-state fermentation. Appl Microbiol Biotechnol 73:533–540

Zhang J, Tang Q, Zimmerman-Kordmann M, Reutter W, Fan H (2002) Activation of B lymphocytes by GLIS, a bioactive proteoglycan from *Ganoderma lucidum*. Life Sci 71:623–638

Chapter 9
Mushrooms as Edible Foods

Gerardo Díaz-Godínez and Maura Téllez-Téllez

9.1 Introduction

Most fungi are multicellular, showing microscopic tubular filaments called "hyphae" that branch and cross over to form the "mycelium" (vegetative phase), which absorb nutrients from their immediate environment. The hyphae of the ascomycete and basidiomycete fungi are divided into cellular cross-walls called septa. A mushroom is the fruiting body or sporocarp, which results from the organization of hyphae where spores are produced, which is why it is the reproductive phase of the fungus. The mushroom has been defined as "a macrofungus with a distinctive fruiting body, which may be hypogeous or epigeous, large enough to be seen with the naked eye and to be picked by hand" (Chang and Miles 1992).

In the fungi species that are considered edible, the fruiting body is usually the material that is consumed. There are many species that are highly appreciated for their flavor and nutritional value, and some mushrooms are considered nutraceuticals because of the benefits they have for the health of those who consume them. There are many species of edible mushrooms that are cultivated and others that are wild. The edible wild mushrooms have been collected and consumed for thousands of years by humans, considering them important in their diet, the archaeological records show the relation that the populations had with their consumption, and with the passage of time, many fungi have been domesticated and produced worldwide. It has been considered that mushrooms are non-timber products of great

G. Díaz-Godínez
Research Center for Biological Sciences, Autonomous University of Tlaxcala,
Tlaxcala, Mexico

M. Téllez-Téllez (✉)
Biological Research Center, Autonomous University of the State of Morelos,
Cuernavaca, Morelos, Mexico
e-mail: maura.tellez@uaem.mx

© The Author(s), under exclusive license to Springer Nature
Switzerland AG 2021
X. Dai et al. (eds.), *Fungi in Sustainable Food Production*, Fungal Biology,
https://doi.org/10.1007/978-3-030-64406-2_9

importance, since they have served as a source of food for many communities, and there is a large number of fungal species that can be used (Boa 2004).

There are historical records of mushrooms that were intentionally cultivated; the first was *Auricularia auricula* around 600 A.D. in China (Chang 1993). The important cultivation of edible mushrooms began around 700 years ago in China with the cultivation of shiitake (*Lentinula edodes*) on inoculated wood logs placed in forests. During the eighteenth century, the cultivation of button mushroom (*Agaricus bisporus*) spread throughout Europe, using flat cultures with mixtures of horse manure placed in caves (Van Griensven 1988). Chang (1993) reported that out of 2000 species that are considered as potentially edible mushrooms, approximately 80 have been experimentally cultivated, 40 of them have been cultivated for economic purposes, about 20 cultivated commercially, and between 5 and 6 have reached an industrial scale in many countries. Most of the mushrooms recognized as cultivated are saprophytic species.

Many species of edible wild mushrooms live in symbiosis with the trees, and this mycorrhizal association supports the growth of natural forests. Therefore, the cultivation of these organisms has been difficult, since the production of fruit bodies depends on a complex set of biotic and abiotic factors. They have a fundamental role in the human diet, agriculture, biotechnology, and forestry. Some of these mushrooms are very appreciated for their flavor and aroma, reaching high selling costs. Of more than 140 species of ectomycorrhizal fungi considered edible, only a small percentage have been grown on a commercial scale (Salerni and Perini 2004). The proposal to use forests for the production of this type of fungus is called mycosilviculture (Savoie and Largeteau 2011).

The demand for mushrooms led to the generation of technology for the domestication of many species, some of which are easy to grow, while others have presented difficulties or have not yet been cultivated, so they are only available by harvest (edible wild mushrooms). It has been estimated that there are more than 1100 mushroom species that are consumed and collected in at least 85 countries (Boa 2004). The harvesting of edible wild mushrooms represents an economic activity in many rural areas of some countries, where they also represent part of their culture since, in addition to being a food, they are often related to rites and traditions. The main cultivated mushrooms are of the genera *Lentinula* 22%, *Pleurotus* 19%, *Auricularia* 18%, *Agaricus* 15%, *Flammulina* 11%, *Volvariella* 5%, and others 10% (Dhar 2017). In Table 9.1 some edible mushrooms are mentioned.

The edible mushrooms are consumed practically all over the world; they are appreciated for their flavor and for their nutritional quality. There are many dishes made with mushrooms where they can be mixed with vegetables, meat, chicken, fish, eggs, etc. and can be eaten fresh or cooked. Of course, the type of mushroom and the dishes depends on the availability of the fungus and on the costumes and traditions of the population. In general, mushrooms are considered a food of excellent nutritional quality, since they have high protein contents (approximately 20–30% dry weight), which are compared with the protein content of milk, eggs, and some meats. It should be mentioned that the amount of mushroom protein depends on the species (and also on the strain) and the growth conditions such as the

Table 9.1 List of some edible mushrooms

Scientific name	Common name	Position in classification (Index Fungorum 2017)	Habitat
Agaricus arvensis Schaeff	Horse mushroom	*Agaricaceae, Agaricales, Agaricomycetidae, Agaricomycetes, Agaricomycotina, Basidiomycota*	Saprobic
Agaricus bisporus (J.E. Lange)	Common mushroom, button mushroom, white mushroom		Saprobic
Agaricus campestris L.	Field mushroom, meadows, pink bottoms		Saprobic
Agaricus comtulus Fr.	None		Saprobic
Agaricus lutosus Møller	None		Saprobic
Agaricus silvaticus Scheff	Pinewood mushroom		Saprobic
Agaricus subrufescens Peck	Almond mushroom		Saprobic
Agrocybe aegerita (V. Brig.) Singer	Black poplar mushroom	*Strophariaceae, Agaricales, Agaricomycetidae, Agaricomycetes, Agaricomycotina, Basidiomycota*	Saprobic
Amanita caesarea (Scop.)	Caesar's mushroom	*Amanitaceae, Agaricales, Agaricomycetidae, Agaricomycetes, Agaricomycotina, Basidiomycota*	Ectomycorrhizal
Amanita codinae (Maire) Bertault	None		Ectomycorrhizal
Amanita constricta Thiers & Ammirati	None		Ectomycorrhizal
Amanita jacksonii Pomerl.	Jackson's slender Caesar		Ectomycorrhizal
Armillaria gallica Marxm. & Romagn.	Brown honey, stumper	*Physalacriaceae, Agaricales, Agaricomycetidae, Agaricomycetes, Agaricomycotina, Basidiomycota*	Saprobic
Armillaria mellea (Vahl) P. Kumm.	Yellow honey, stumpers		Parasitic/ saprobic
Auricularia auricula-judae (Bull.) Quél.	Wood ears	*Auriculariaceae, Auriculariales,* Incertae sedis, *Agaricomycetes, Agaricomycotina, Basidiomycota*	Saprobic
Auricularia mesentérica (Dicks.) Pers	None		Saprobic
Auricularia polytricha (Mont.) Sacc.	None		Saprobic

(continued)

Table 9.1 (continued)

Scientific name	Common name	Position in classification (Index Fungorum 2017)	Habitat
Boletus aereus Bull.	Bronze bolete	*Boletaceae, Boletales, Agaricomycetidae, Agaricomycetes, Agaricomycotina, Basidiomycota*	Ectomycorrhizal
Boletus badius (Fr.) Fr.	Bay bolete		Ectomycorrhizal
Boletus chrysenteron Bull.	Red cracked bolete		Ectomycorrhizal
Boletus edulis Bull.	King bolete		Ectomycorrhizal
Boletus pinicola Rea	Pine bolete, pinewood king bolete		Ectomycorrhizal
Boletus reticulatus Schaeff.	Summer cep		Ectomycorrhizal
Calocybe gambosa (Fr.) Singer	St. George's mushroom	*Lyophyllaceae, Agaricales, Agaricomycetidae, Agaricomycetes, Agaricomycotina, Basidiomycota*	Saprobic
Calvatia gigantea (Batsch) Lloyd	Giant puffball	*Agaricaceae, Agaricales, Agaricomycetidae, Agaricomycetes, Agaricomycotina, Basidiomycota*	Saprobic
Calvatia utriformis (Bull.) Jaap	Mosaic puffball		Saprobic
Clavariadelphus truncatus Donk	Club coral	*Clavariadelphaceae, Gomphales, Phallomycetidae, Agaricomycetes, Agaricomycotina, Basidiomycota*	Ectomycorrhizal
Cantharellus cibarius Fr.	Girolle, chanterelle	*Cantharellaceae, Cantharellales*, Incertae sedis, *Agaricomycetes, Agaricomycotina, Basidiomycota*	Ectomycorrhizal
Cantharellus cinnabarinus (Schwein.) Schwein.	Red chanterelle, cinnabar-red chanterelle		Ectomycorrhizal
Cantharellus cinereus (Pers.) Fr.	*Merulius cinereus*		Ectomycorrhizal
Coprinus comatus (O.F. Müll.) Pers.	Shaggy manes, shaggy ink cap, lawyer's wig	*Agaricaceae, Agaricales, Agaricomycetidae, Agaricomycetes, Agaricomycotina, Basidiomycota*	Saprobic
Cortinarius praestans (Cordier) Gillet	Goliath webcap	Cortinariaceae, *Agaricales, Agaricomycetidae, Agaricomycetes, Agaricomycotina, Basidiomycota*	Ectomycorrhizal
Cortinarius purpurascens Fr.	None		Ectomycorrhizal
Cortinarius variicolor (Pers.) Fr.	Contrary webcap		Ectomycorrhizal

(continued)

Table 9.1 (continued)

Scientific name	Common name	Position in classification (Index Fungorum 2017)	Habitat
Craterellus cornucopioides (L.) Pers.	Black trumpets, black chanterelle	*Cantharellaceae, Cantharellales*, Incertae sedis, *Agaricomycetes, Agaricomycotina, Basidiomycota*	Ectomycorrhizal
Craterellus fallax A.H. Sm.	Black trumpets		Ectomycorrhizal
Craterellus tubaeformis (Fr.) Quél.	Yellowfoot, winter mushroom, funnel chanterelle		Ectomycorrhizal
Cyttaria espinosae Lloyd	Digüeñe	*Cyttariaceae, Cyttariales, Leotiomycetidae, Leotiomycetes, Pezizomycotina, Ascomycota*	Parasitic
Chlorophyllum rhacodes (Vittadini) Vellinga	Shaggy parasol	*Agaricaceae, Agaricales, Agaricomycetidae, Agaricomycetes, Agaricomycotina, Basidiomycota*	Saprobic
Chroogomphus rutilus (Schaeff.) O.K. Mill.	Pine-spikes, spike-caps	*Gomphidiaceae, Boletales, Agaricomycetidae, Agaricomycetes, Agaricomycotina, Basidiomycota*	Ectomycorrhizal
Entoloma abortivum (Berk. & M.A. Curtis) Donk	Aborted entoloma	Entolomataceae, *Agaricales, Agaricomycetidae, Agaricomycetes, Agaricomycotina, Basidiomycota*	Ectomycorrhizal
Fistulina hepatica (Schaeff.) With.	Beefsteak polypore, the ox tongue	*Fistulinaceae, Agaricales, Agaricomycetidae, Agaricomycetes, Agaricomycotina, Basidiomycota*	Saprobic
Flammulina velutipes (Curtis) Singer	Velvet shank, winter fungus, velvet foots	*Physalacriaceae, Agaricales, Agaricomycetidae, Agaricomycetes, Agaricomycotina, Basidiomycota*	Saprobic
Grifola frondosa (Dicks.) Gray	Hen of the woods	*Meripilaceae, Polyporales*, Incertae sedis, *Agaricomycetes, Agaricomycotina, Basidiomycota*	Saprobic
Gyroporus castaneus (Bull.) Quél.	Chestnut bolete	*Gyroporaceae, Boletales, Agaricomycetidae, Agaricomycetes, Agaricomycotina, Basidiomycota*	Ectomycorrhizal

(continued)

Table 9.1 (continued)

Scientific name	Common name	Position in classification (Index Fungorum 2017)	Habitat
Hericium coralloides (Scop.) Pers.	Coral tooth fungus	*Hericiaceae, Russulales,* Incertae sedis, *Agaricomycetes, Agaricomycotina, Basidiomycota*	Saprobic
Hericium erinaceus (Bull.) Pers.	Lion's mane, bear's tooth		Saprobic
Hydnum repandum L.	Sweet tooth	*Hydnaceae, Cantharellales,* Incertae sedis, *Agaricomycetes, Agaricomycotina, Basidiomycota*	Ectomycorrhizal
Hydnum rufescens Pers.	The terracotta hedgehog		Ectomycorrhizal
Hydnum umbilicatum Peck.	Depressed hedgehog		Ectomycorrhizal
Hygrophorus chrysodon (Batsch) Fr.	Gold-flecked woodwax	*Hygrophoraceae, Agaricales, Agaricomycetidae, Agaricomycetes, Agaricomycotina, Basidiomycota*	Ectomycorrhizal
Hygrophorus latitabundus Britzelm.	Gray llenega, black llenega		Ectomycorrhizal
Hygrophorus russula (Schaeff. ex Fr.) Kauffman	False russula and russula-like waxy cap		Ectomycorrhizal
Hypomyces lactifluorum (Schwein.) Tul. & C. Tul.	Lobster mushroom	*Hypocreaceae, Hypocreales, Hypocreomycetidae, Sordariomycetes, Pezizomycotina, Ascomycota*	Parasitic
Laccaria ochropurpurea (Berk.) Peck	Purple-gilled Laccaria	*Hydnangiaceae, Agaricales, Agaricomycetidae, Agaricomycetes, Agaricomycotina, Basidiomycota*	Ectomycorrhizal
Lactarius deliciosus (L.) Gray	Orange-latex milky	*Russulaceae, Russulales,* Incertae sedis, *Agaricomycetes, Agaricomycotina, Basidiomycota*	Ectomycorrhizal
Lactarius deterrimus Gröger	Orange milkcap		Ectomycorrhizal
Lactarius salmonicolor R. Heim & Leclair	Milky agaric		Ectomycorrhizal
Lactarius subdulcis (Pers.) Gray	Mild milkcap		Ectomycorrhizal
Lactarius volemus (Fr.) Fr.	Weeping milkcap, voluminous-latex milky		Ectomycorrhizal

(continued)

Table 9.1 (continued)

Scientific name	Common name	Position in classification (Index Fungorum 2017)	Habitat
Laetiporus cincinnatus (Morgan) Burds., Banik & T.J. Volk	Chicken of the woods	*Fomitopsidaceae, Polyporales,* Incertae sedis, *Agaricomycetes, Agaricomycotina, Basidiomycota*	Parasitic/ saprobic
Laetiporus sulphureus (Bull.) Murrill	Chicken of the woods, chicken mushroom		Parasitic/ saprobic
Leccinum aurantiacum (Bull.) Gray	Red-capped scaber stalk	*Boletaceae, Boletales, Agaricomycetidae, Agaricomycetes, Agaricomycotina, Basidiomycota*	Ectomycorrhizal
Leccinum scabrum (Bull.) Gray	Birch bolete		Ectomycorrhizal
Leccinum versipelle (Fr. & Hök) Snell	Orange birch bolete, boletus testaceoscaber		Ectomycorrhizal
Lentinula edodes (Berk.) Pegler	Shiitake	*Omphalotaceae, Agaricales, Agaricomycetidae, Agaricomycetes, Agaricomycotina, Basidiomycota*	Saprobic
Lepiota americana (Peck) Sacc.	Reddening lepiota	*Agaricaceae, Agaricales, Agaricomycetidae, Agaricomycetes, Agaricomycotina, Basidiomycota*	Saprobic
Lepista nuda (Bull.) Cooke	Blewits	Tricholomataceae, *Agaricales, Agaricomycetidae, Agaricomycetes, Agaricomycotina, Basidiomycota*	Saprobic
Leucoagaricus leucothites Vittad. Wasser	White dapperling, white agaricus mushroom	*Agaricaceae, Agaricales, Agaricomycetidue, Agaricomycetes, Agaricomycotina, Basidiomycota*	Saprobic
Lycoperdon umbrinum Pers.	Common puffball, warted puffball, gem-studded puffball, the devil's snuffbox		Saprobic
Macrolepiota procera (Scop.) Singer	Parasol mushroom		Saprobic
Macrolepiota rhacodes (Vittad.) Singer	Shaggy parasol		Saprobic

(continued)

Table 9.1 (continued)

Scientific name	Common name	Position in classification (Index Fungorum 2017)	Habitat
Marasmius oreades (Bolton) Fr.	Fairy ring champignon	*Marasmiaceae, Agaricales, Agaricomycetidae, Agaricomycetes, Agaricomycotina, Basidiomycota*	Saprobic
Phylloporus rhodoxanthus (Schwein.) Bres.	Gilled bolete	*Boletaceae, Boletales, Agaricomycetidae, Agaricomycetes, Agaricomycotina, Basidiomycota*	Ectomycorrhizal
Pleurotus citrinopileatus Singer	Golden oyster mushroom	Pleurotaceae, *Agaricales, Agaricomycetidae, Agaricomycetes, Agaricomycotina, Basidiomycota*	Saprobic
Pleurotus cystidiosus O.K. Mill	Abalone mushroom		Saprobic
Pleurotus djamor var. *roseus* Corner, Beihefte	Pink oyster mushroom		Saprobic
Pleurotus eryngii (DC.) Quél.	King oyster		Saprobic
Pleurotus ostreatus (Jacq.) P. Kumm.	Oyster mushroom		Saprobic
Pleurotus pulmonarius (Fr.) Quél.	Indian oyster, Italian oyster, phoenix mushroom, lung oyster		Saprobic
Pleurotus porrigens (Pers.) P. Kumm.	Angel wings		Saprobic
Pleurotus sajor-caju (Fr.) Singer	Gray oyster mushroom, phoenix-tail mushroom		Saprobic
Pleurotus sapidus Sacc.	Black oyster mushrooms		Saprobic
Pluteus cervinus (Schaeff.) P. Kumm.	Deer or fawn mushroom	Pluteaceae, *Agaricales, Agaricomycetidae, Agaricomycetes, Agaricomycotina, Basidiomycota*	Saprobic
Ramaria aurea (Schaeff.) Quél.	Golden coral	*Gomphaceae, Gomphales, Phallomycetidae, Agaricomycetes, Agaricomycotina, Basidiomycota*	Ectomycorrhizal

(continued)

Table 9.1 (continued)

Scientific name	Common name	Position in classification (Index Fungorum 2017)	Habitat
Ramaria botrytis (Pers.) Bourdot.	Clustered coral, the pink-tipped coral mushroom, the cauliflower coral	*Gomphaceae, Gomphales, Phallomycetidae, Agaricomycetes, Agaricomycotina, Basidiomycota*	Ectomycorrhizal
Rhizopogon luteolus Fr.	Yellow false truffle	*Rhizopogonaceae, Boletales, Agaricomycetidae, Agaricomycetes, Agaricomycotina, Basidiomycota*	
Russula virescens (Schaeff.) Fr	Green quilted russula	*Russulaceae, Russulales,* Incertae sedis, *Agaricomycetes,* *Agaricomycotina, Basidiomycota*	Ectomycorrhizal
Russula cyanoxantha (Schaeff.) Fr.	Charcoal burner		Ectomycorrhizal
Russula heterophylla (Fr.) Fr.			Ectomycorrhizal
Russula olivacea (Schaeff.) Fr.	Olive brittlegill		Ectomycorrhizal
Russula vesca (Fries)	Bare-toothed russula, the flirt		Ectomycorrhizal
Russula virescens (Schaeff.) Fr.	Green-cracking russula, the quilted green russula, green brittlegill		Ectomycorrhizal
Sparassis crispa (Wulfen) Fr.	Cauliflower mushroom	*Sparassidaceae, Polyporales,* Incertae sedis, *Agaricomycetes, Agaricomycotina, Basidiomycota*	Saprobic
Suillus americanus (Peck) Snell	Chicken fat	*Suillaceae, Boletales, Agaricomycetidae, Agaricomycetes, Agaricomycotina, Basidiomycota*	Ectomycorrhizal
Suillus bovinus (L.) Roussel	Jersey cow mushroom, bovine bolete		Ectomycorrhizal
Suillus brevipes (Peck) Kuntze	Short-stalked suillus		Ectomycorrhizal
Suillus granulatus (L.) Roussel	Weeping bolete, granulated bolete		Ectomycorrhizal
Suillus luteus (L.) Roussel	Slippery jack		Ectomycorrhizal
Suillus tomentosus Singer	Blue-staining slippery jack, poor man's slippery jack, woolly capped		Ectomycorrhizal
Suillus variegatus (Sw.) Richon & Roze	Velvet bolete, variegated bolete		Ectomycorrhizal

(continued)

Table 9.1 (continued)

Scientific name	Common name	Position in classification (Index Fungorum 2017)	Habitat
Tricholoma matsutake (S. Ito & S. Imai) Singer	Matsutake	Tricholomataceae, *Agaricales*, *Agaricomycetidae*, *Agaricomycetes*, *Agaricomycotina*, *Basidiomycota*	Ectomycorrhizal
Tricholoma flavovirens (Pers.) S. Lundell	Man on horseback, canary trich		Ectomycorrhizal
Tricholoma terreum (Schaeff.) P. Kumm.	Gray knight, dirty tricholoma		Ectomycorrhizal
Ustilago maydis (DC.) Corda	Delicacy huitlacoche	Ustilaginaceae, Ustilaginales, Ustilaginomycetidae, Ustilaginomycetes, Ustilaginomycotina, Basidiomycota	Parasitic
Volvariella volvacea (Bull.) Singer	Paddy straw mushroom	Pluteaceae, *Agaricales*, *Agaricomycetidae*, *Agaricomycetes*, *Agaricomycotina*, *Basidiomycota*	Saprobic

substrate, pH, temperature, and humidity among others. Mushrooms are good sources of almost all essential amino acids when compared with common vegetables; besides their digestibility is very good. On average, the moisture content of the mushrooms is 90% in fresh weight. In the dry matter, carbohydrates are the major components with values of approximately 40% constituted by glycogen, chitin which is the main component of the fiber, glucan, chitosan, and mannans (simple carbohydrates are found in low quantities, since fungi do not perform photosynthesis).

Mushrooms provide an average of 85–125 kJ/100 g, while an adult male needs about 10,000 kJ per day. This low energy value of mushrooms enables it to be used in low-calorie diets which are ideal for diabetics (Moore 2017). Another very important characteristic is that the fungi are free of cholesterol, so its consumption helps reduce the risk of coronary heart disease and other related disorders, also present very low lipid content (2–8% dry weight), constituted by triglycerides, free fatty acids, glycerides, sterols, and phospholipids. The mushrooms are an excellent source of B-complex vitamins (riboflavin (B2), niacin, pantothenic acid, thiamin (B1), biotin, folate, and vitamin B12) and vitamin D and also of minerals such as copper, selenium, potassium, phosphorous, zinc, and magnesium. Mushrooms are important and necessary in the diet due to their nutritional quality and the health benefits they provide such as blood pressure control, lowering of blood cholesterol, glucose regulation in blood, regulation of the intestinal function, control of tumor growth, anti-inflammatory effect, and antioxidant activity among others (Téllez-Téllez and Díaz-Godínez 2017).

9.2   Types of Mushrooms

Mushrooms can be classified into three groups: the saprophytes, the parasites, and the mycorrhizae. Most of the gourmet mushrooms are saprophytic (wood-decomposing fungi, which are the premier recyclers on the planet). The enzymes and acids that they secrete degrade large molecular complexes into simpler compounds. All ecosystems depend on the ability of fungi to decompose the organic matter of plants; the final result of their activity is the return of carbon, hydrogen, nitrogen, and minerals to the ecosystem in usable forms. Many saprophytic fungi can be weakly parasitic in their behavior. Some parasitic mushrooms are edible; the best known is *Armillaria mellea* (Stamets 2005). Mycorrhizae are systems of fungal symbiosis with the roots of plants, where both species benefit. In this case, the plant receives mainly mineral nutrients and water from the fungus, and the fungus obtains vitamins and carbohydrates produced by photosynthesis from the plant. It is estimated that a high percentage of terrestrial plants present mycorrhizas on a regular basis.

9.2.1   Saprophytic Mushrooms

Agaricus subrufescens **Peck**   It belongs to the *Agaricaceae* family of the order *Agaricales* of the basidiomycetes (Index Fungorum 2017). This species and others of the same genus can be found in forests, but more commonly in pastures or roadsides; it is a saprophyte that uses decaying leaves often at the borders between forests and parks (Firenzuoli et al. 2008). It was called the "almond mushroom," because of its almond flavor. It is cultivated and consumed in the United States from the late nineteenth century to the early twentieth century. In 1960, it was discovered in Brazil and called the "Piedade mushroom" because of the name of the town where it was found, in the province of São Paulo (Kerrigan 2005).

The champignon mushroom has been produced in a commercial market on a national scale in Brazil since the early 1990s and has been exported to other countries, currently also grown at industrial level in Japan, China, Taiwan, and Korea. As substrates for the production of fruiting bodies, it has been used as sugarcane bagasse, some herbs (*Brachiaria* spp., *Cynodon dactylon*, *Panicum maximum*), cereal straw (*Triticum aestivum*, *Avena sativa*, *Oryza sativa*), and manure supplemented with nitrogen sources (soy, wheat, cotton flour, urea, ammonium sulfate) and sources of phosphorus and calcium (Llarena-Hernández et al. 2014). The fruiting bodies of *Agaricus subrufescens* contain 89–91% water and have high level of minerals (potassium, phosphorus, calcium, magnesium, and zinc). However, small amounts of cadmium have been detected (Gyorfi et al. 2010). The polysaccharides consisted of 57.7% glucose, 27.7% galactose, 7.3% mannose/xylose, and 4% fucose (Volman et al. 2010).

Agrocybe aegerita (**V. Brig.**) **Singer**   This mushroom is a white-rot basidiomycete, commercially cultivated in Italy and highly appreciated as a delicacy (so-called Pioppino mushroom) (Ullrich et al. 2004). The life cycle of *Agrocybe aegerita* is controlled by the tetrapolar mechanism of homogeneous incompatibility that involves the two unlinked factors A and B. When it has favorable environmental conditions, the dikaryon can differentiate the fruiting bodies (basidiocarps); subsequently karyogamy and meiosis are carried out (Meinhardt and Leslie 1982). The common name for the mushroom is black poplar mushroom or chestnut mushroom, since it is often found on poplar wood logs, growing saprophytically, often clusters, on stumps in the southeastern United States and southern Europe, in similar climatic zones of the Far East. It has fruited this species on supplemented oak and alder sawdust/chip; willow, poplar, and maple are just as likely to support substantial fruitings. It is one of the most cultivated mushrooms in Asia and has been highly valued as a functional food for its antitumor, antioxidant, antifungal, hypocholesterolemic, and hypolipidemic effects (Diyabalanage et al. 2008).

Auricularia auricula-judae (**Bull.**) **Quél.**   It is a heterobasidiomycete of the *Auriculariales* order and the *Auriculariaceae* family (Index Fungorum 2017); common names include Jew's ear, wood ear, jelly ear, and wood-ear mushrooms. It is cultivated in China, Taiwan, Thailand, the Philippines, Indonesia, and Malaysia (Tang et al. 2010). The fruiting body has a waxy and cartilaginous texture, the color varies from purplish brown to black, especially when dried, its horizontally septed basidium makes them significantly different in taste and morphological characters (somewhat variable irregular forms, from open to sometimes deformed) of other cultivated mushrooms, such as *Agaricus bisporus* and *Pleurotus* spp., among others.

Today, *Auricularia* mushrooms are among the four most important cultivated mushrooms in the world, mainly in China and Southeast Asia, with an annual world production of 420,000 tons. They can grow well on various organic agricultural and industrial wastes (Yan et al. 2004). *Auricularia auricula-judae* is a culinary-medicinal mushroom, which contains potassium (172.03 mg/kg) and manganese (1.66 mg/kg). Glutamic acid has been detected in high amounts (10.09 mg/kg); it also contains cysteine (0.34 mg/kg) and methionine (0.80 mg/kg). The predominant oil is the methyl ester of 9,12-octadecadienoic acid (Ohiri and Bassey 2017). The fragrance was described as musty, and reminiscent of raw compost (Stamets 2000).

Coprinus comatus (**O.F. Müll.**) **Pers.**   It is a member of the family *Agaricaceae* and order *Agaricales*, commonly called as chicken drumstick mushroom, shaggy ink cap, lawyer's wig, or shaggy mane. *Agaricales* are common mushrooms that are often seen on lawns, gravel roads, and waste areas around the world (Park and Lee 2005), which grow in late summer and fall throughout the temperate region of the world, and its smell has been described as farinaceous and mildly sweet (Stamets 2000).

For many years, the Chinese population has appreciated this species for its high nutritional value and its delicious flavor. Its fruiting body is white, it is thin and fragile, the stipe is something hard and fibrous, and as it matures it changes color to pink and then to black and exudes a type of ink of that color. Today, these mushrooms are widely cultivated in China and their production exceeds 380,000 tons per year (CEFA 2008). The substrates used for cultivation include cotton waste, corncobs, rice straw, urea, steer manure, lime, etc. (Yang and Xue 2000). Stojković et al. (2013) reported the nutrient content of wild and cultivated species, among which there was no significant difference in carbohydrate content (76% dry weight), ash (10%), and energy (368 kcal/100 g), but differences in protein (wild 10.9%; cultivated 11.84%) and fat (wild 1.8%; cultivated 1.98%).

Flammulina velutipes (**Curtis**) **Singer**   It is a basidiomycete that belongs to the family *Physalacriaceae* (Index Fungorum 2017). "Enokitake" is the Japanese name with which it is known, which is also called as golden needle mushroom or winter mushroom. It is a saprophyte, mainly on hardwoods, occasionally on conifers, commonly growing from sea level to tree line (Stamets 2000), mushroom of great popularity in Asia, requires low temperature to develop (Chang and Miles 2004). It has been reported that its morphological characteristics are influenced by the incidence of light and temperature (Sakamoto et al. 2007). *Flammulina velutipes* is one of the six most actively cultivated mushroom species in the world; more than 300,000 tons of this mushroom are produced per year (Psurtseva 2005). Because it is a white-rot fungus, it can be found in dead wood. In the commercial cultivation, it has been used mainly as wood (sawdust), alone or mixed with ears of corn, cottonseed husk, bagasse of sugarcane, etc. in cold treatment (Royse 1995). The reduction of the cultivation cycles, the cold treatment, and the amount of waste material are determining factors in its cultivation. Most strains require a cold shock and/or a low growth temperature (4.4–13 °C), which vary in their sensitivity to light and carbon dioxide levels (Stamets 2000).

Grifola frondosa (**Dicks.**) **Gray**   *Grifola frondosa* is commonly called as hen of the woods and sheep's head, but maitake is its Japanese name meaning "dancing mushroom" derived from folklore, because when a person stumbles upon the mushroom, he might dance out of happiness because of the mushroom's high market value. This mushroom has a unique appearance. It is fleshy polyporaceous recognized by its undulating pileus and smoky brown color; it is organized in large clusters forming rosettes that emerge from a single branched structure of the stipe. The mature fruiting body is fleshy dark gray-brown and becomes light gray with age (Chang and Miles 2004). It is a saprophyte mushroom that grows on trunks of dead, dying, or aged woods, such as oak, elm, maple, and chestnut (Chen et al. 2000). The first large-scale commercial production was developed in Japan in 1981; 10 years later the United States and China did. In 1999, Japanese growers produced almost 40,000 tons, and in 2001, China produced 14,600 tons. It has been reported that this fungus contains 19.7% of crude protein and 13. 1% of pure protein, 61.06% of carbohydrates, 9.7% of fiber, and 3.2% of lipids (Chang and Miles 2004).

Hericium erinaceus (**Bull.**) **Pers.** *Hericium erinaceus* known as yamabushitake (Japanese), lion's mane, monkey's mushroom, bear's head, white beard, hog's head fungus, old man's beard, pom-pom, and bearded tooth (Thongbai et al. 2015). It is within the family *Hericiaceae*, order *Russulales*, and class *Agaricomycetes* (Index Fungorum 2017). It is considered as a saprophyte or weak parasite, since it has been found in dead wood, knots, or cracks of living hard hardwoods. The mature fruiting body is fleshy and semispherical of whitish color and gradually changes from yellowish to brownish (He et al. 2017).

During 1959–1960 this fungus was only available in the wild, because it grew in the deep forest on old and dead wood. The macromorphology of *Hericium erinaceus* is usually sufficient for identification. However, it is not easy to differentiate it from *Hericium coralloides*, since the sizes of basidiospores are very similar in both species, but it has been reported that the fruiting bodies of *Hericium coralloides* tend to be much more branched. So their substrates have been used to aid identification, since *Hericium coralloides* is associated with conifers, while *Hericium erinaceus* is deciduous trees (Stamets 2005). It was domesticated by scientists from the Shanghai Agricultural Academy of Science and later distributed to other places. The fungus grows on different substrates, including artificial media, also cheap substrates such as agro-industrial waste (sugarcane bagasse, sawdust, cottonseed hull, and corncobs), and tofu, cheese, and whey. The fruiting body presents 18.8% protein, 61.3% carbohydrates, 2.9% lipids, and 6.91% ash (Rodrigues et al. 2015).

Pleurotus **spp.** The cultivation of *Pleurotus* increased rapidly, several species of this genus are found all over the world, and their edibility is highly appreciated. It is cultivated in many countries including Italy, Germany, Belgium, China, Japan, Taiwan, India, Singapore, Thailand, Pakistan, and Indonesia. Beginning in 1997, world production increased; cultivation began in Nigeria, Mexico, Brazil, Colombia, Canada, the United States, Tanzania, Zambia, and Malawi. The great interest in these species is due to their flavor, nutritional content, and the wide range of enzymes that they secrete, which allows them to degrade all the polysaccharides found in biomass of forest and agriculture (cellulose, hemicellulose, and lignin), so production costs are more accessible (Chang and Miles 2004).

The genus *Pleurotus* gathers several species, such as *Pleurotus ostreatus* (Jacq.) P. Kumm (oyster mushroom), *Pleurotus sajor-caju* (Fr.) Singer (gray oyster mushroom or phoenix-tail mushroom), *Pleurotus citrinopileatus* Singer (golden oyster mushroom), *Pleurotus cystidiosus* O.K. Miller (abalone mushroom, maple oyster), *Pleurotus djamor* var. *roseus* Corner, Beihefte (pink oyster mushroom), *Pleurotus sapidus* Sacc. Syll. fung. (black oyster mushrooms), *Pleurotus pulmonarius* (Fr.) Quél. (Indian oyster, Italian oyster, phoenix mushroom, lung oyster), and *Pleurotus eryngii* (DC.) Quél. (king oyster). They belong to the family Pleurotaceae, order *Agaricales*, and class *Basidiomycota* (Index Fungorum 2017). They grow naturally in temperate zones or in colder seasons in subtropical areas in rotten trees such as oak, elm, maple, and poplar, among others (Chang and Miles 2004).

World production of *Pleurotus* was around 3.4 × 10⁶ tons in 2008, with China being the country with the highest production (1.5 × 10⁶ tons) followed by the United States (0.38 × 10⁶ tons) (Melo de Carvalho et al. 2010). For its cultivation, different types of substrates have been used, such as agricultural residues (corn, wheat, barley, lentils, rice husks) and plant residues such as coffee pulp, sawdust, and sugarcane bagasse, among others. Cultivation of *Pleurotus* represents an excellent alternative to recycle several residues, helping to reduce the presence of these materials in the environment (Pandey et al. 2000). The fruiting bodies contain, on average, 25% protein (contains all the essential amino acids), 63% carbohydrates, 14% crude fiber, and 4% lipids (Chang and Miles 2004). Grain spawn of *Pleurotus citrinopileatus* smells astringent, acrid, nutty, and sometimes fishy, with a scent that, in time, is distinctly recognizable to this species. *Pleurotus cystidiosus* smells musty, farinaceous not anise-like, and for *Pleurotus ostreatus*, its smell is sweet, rich, pleasant, distinctly anise, and almost-like. Grain spawn of *Pleurotus pulmonarius* smells sweet, pleasant, and distinctly "oyster-esque" (Stamets 2000).

Volvariella volvacea **(Bull.) Singer** Known as "straw mushroom" or "Chinese mushroom," it belongs to the family Pluteaceae of the basidiomycetes (Index Fungorum 2017). It is a high-temperature fleshy saprophyte fungus, it grows in tropical and subtropical regions of Asia, and it can also grow seasonally in temperate regions. It has been described as a homothallic basidiomycete, so the homokaryotic mycelium arising from germination is able to convert to the dikaryotic form and complete the cycle; the dikaryotic hyphae lack pincer connections, the morphological markers that differentiate the dikaryon from the homokaryon (Chang and Ling 1970). In the eighteenth century, straw cultivation was initiated by Buddhist monks from the Nanhua Temple located in Guangdong Province. Their diet was enriched by cultivating the fungus on fermented rice straw; the mushroom was a tribute to Chinese royalty (Chang 1977).

Volvariella volvacea has been cultivated using cotton waste, rice straw, water lily, sugarcane bagasse, palm oil residues, and wastes of coconut, of banana, and of pineapple, among others. The best characteristics of the cotton waste compost have led to the cultivation of the mushroom to a semi-industrialized process in Hong Kong, Taiwan, Indonesia, China, and Thailand (Chang and Miles 2004). Annual production has increased in recent years; in 2010 the mushroom production in China was 330,000 tons, which represents more than 80% of the world production (Bao et al. 2013).

9.2.2 Mycorrhizal Mushrooms

Tuber **spp.** *Tuber* spp. are edible mushrooms of hypogeous growth (under the ground) that establish a symbiotic relationship (ectomycorrhiza) with various species of gymnosperms and angiosperms in a variety of habitats (subtropical cloud forests, temperate forests, boreal forests, among others) (Bulman et al. 2010). *Tuber* spp. are edible mushrooms of hypogeous growth (under the ground) that establish a symbiotic relationship (ectomycorrhiza) with various species of gymnosperms and angiosperms in a variety of habitats (subtropical cloud forests, temperate forests, boreal forests, among others) (Gioacchini et al. 2005). It has been reported that there are different biotic (fungi, yeast, bacteria, mesofauna, plant host) and abiotic factors (soil composition, humidity, temperature, pH, etc.) involved in the formation of fruiting bodies (Ceruti et al. 2003). *Tuber melanosporum* is collected in the south and Western Europe (Italy, France, and Spain), and *Tuber magnatum* has been collected in Italy and in Eastern Europe (Croatia, Slovenia, and Hungary), from September to January; truffles are sought with the aid of pigs or dogs trained (Mello et al. 2006).

The price of the kilogram of *Tuber melanosporum* in Europe ranges from 150 to 800 euro (farmers) and to retailers is higher, for example, in Paris and London, it is 2000–4000 euros and in Australia, in 2012 it was 950 euros (Duell 2012). Given the economic importance, for the first time in France during the nineteenth century, the semi-cultivation of these species began (Olivier et al. 1996). Currently, cultivation processes are being implemented throughout the world (mainly in the Mediterranean climate). Despite research efforts, the mushroom cultivation is not completely domesticated (Wang 2012), but *Tuber melanosporum* genome sequencing can be used to improve the results (Martin et al. 2010).

Tricholoma matsutake (**S. Ito & S. Imai**) **Singer** It is known as "matsutake." It is an ectomycorrhizal basidiomycete that grows mainly in Japanese red pine (*Pinus densiflora*) in forests in Japan; it belongs to the family Tricholomataceae and order *Agaricales* (Index Fungorum 2017). This species forms underground mycelial aggregations called "shiros," which usually form a circular arrangement of fruiting bodies known as fruitful fairy rings (Ogawa et al. 1978). Within its life cycle, it alternates a monokaryon-dikaryon, which is typical of majority of basidiomycetes. The fruiting bodies of "matsutake" are economically important edible fungi in Japan; to meet domestic demand, approximately 3000 tons of these fungi are imported annually, mainly from Korea, China, North America, and North Africa. In the Japanese market, 1 kg of top quality fresh matsutake can be sold for up to $2000 US (Gill et al. 2000). Due to the varying ecological conditions, mainly precipitation and temperature, the natural production of the fungus is limited and difficult to forecast (Amend et al. 2009). Efforts have been made to establish an artificial cultivation system that meets the high-quality annual demand (Ogawa et al. 1978; Tominaga 1978; Yamada et al. 1999; Guerin-Laguette et al. 2000). Unfortunately, it has not yet been successful, so the research seeks to know the factors that influence the development of the fruitful body; the relationship of ecological factors such as soil quality, climate, and the host plant; and even the associated microbial communities, in order to optimize the cultivation process.

Amanita caesarea (**Scop.**) **Pers.** It is a delicious edible species known as Caesar's mushroom (Yun and Hall 2004). In many countries the edible fungus *Amanita caesarea* is of great social and economic importance, due to its high gastronomic value, which has been appreciated since Roman times (Sitta and Davoli 2012). It grows solitary and sometimes in groups in deciduous forests. This ectomycorrhizogenic fungus is symbiotically associated with *Pinus strobus*, *Pinus virginiana*, *Castanopsis carlesii*, *Castanopsis hystrix*, *Castanea sativa*, *Quercus baronii*, *Quercus faginea*, *Quercus robur, and Quercus suber*, among others (González et al. 2002). It has been reported to grow in temperate and tropical regions of Africa, Asia, Australia, Europe, and Central and North America, but not in South America (Sánchez-Ramírez et al. 2015). It has a distinctive orange cap, yellow gills, and stem. *Amanita caesarea* is a thermophilic species and grows in different soil texture with pH between 6 and 6.5 (Daza et al. 2006). It has been worked to obtain a greater number of fruiting bodies; it has been reported that within the symbiosis between the fungus and the root, bacteria that promote the establishment of this symbiosis have also been implicated. Cano et al. (2017) reported that using bacterial strains (*Pseudomonas fluorescens* and *Bacillus cereus*) that promote mycorrhization under in vitro conditions, it was observed that *Pseudomonas fluorescens* showed spatial and temporal compatibility with *A. caesarea* that did not affect the growth of the fungus, but *Bacillus cereus* did not show compatibility and the growth of the fungus was a little affected. So they are still looking for bacteria that can improve the symbiosis process.

Boletus edulis **(Bull.) Fr.** The genus *Boletus* is cosmopolitan, widely found in temperate zones such as the northern and southern hemispheres, inhabiting tropical and middle latitude forests. Common names are cep and penny bun mushroom (English), cépe de Bordeaux (French), Steinpilz (German), porcino (Itahan), zhutm mo (pig leg mushroom, North China), and dajiao gu (fat feet mushroom, South China) (Hall et al. 1998). The *Boletus edulis* species complex (*Boletus edulis* sensu lato) is an edible ectomycorrhizal fungus with excellent culinary qualities including four European species (*Boletus edulis*, *Boletus aereus* Bull. Fr., *Boletus pinophilus* Pilát and Dermek, and *Boletus aestivalis* Fr.) (Leonardi et al. 2005). Unlike most other mushrooms, *Boletus edulis* can be stored dry and cooked at high temperatures, without a significant loss of its distinctive flavor (Ney and Freytag 1980). The flavor of this dried mushroom including odor and taste is marvelous nutty, earthy, and meaty all at once; for this reason it has the great demand within the international gastronomy (Tsai et al. 2007). In the last 30 years, there has been a strong increase in the demand for edible mushrooms. To meet the demand, large quantities of mushrooms were imported to Italy, mainly from Eastern European countries (Yugoslavia, Romania, Bulgaria, Poland) and Asia. Consequently, commercial harvesting for the domestic and export markets becomes increasingly important (Arnolds 1995). It is not known exactly the net value of consumption or that which is used as an ingredient in food, it has been estimated 20,000–100,000 metric tons can be consumed annually, and the price to wholesalers of the fresh fruiting bodies in the United States in 2009 was between US $ 60 and 200 per kg (Dentinger et al. 2010).

Work is being done to increase the production of fruiting bodies (Poitou et al. 1982; Duñabeitia et al. 1996; Meotto et al. 1999; Yamanaka et al. 2000). Monitoring of environmental conditions in the *Boletus* species was carried out for 10 years; it was observed that the main factor inducing fructification is precipitation, which leads to a 40% increase in soil moisture; the production of the fruiting bodies can vary from 6 to 25 days, depending on the temperature of the soil. For *Boletus edulis*, a cold shock of 5 °C is necessary. Water and temperature vary with fungal species (Savoie and Largeteau 2011).

9.2.3 Parasitic Mushrooms

Armillaria mellea (Vahl) P. Kumm Basidiomycete (honey mushroom) is edible and medicinal belonging to the family *Physalacriaceae* and order *Agaricales* (Index Fungorum 2017); they have described it as a parasite or saprophyte in many hosts of woody (conifer) or broadleaf plants from spring to autumn (Park and Lee 1991). It is known that the fruiting body of *Armillaria mellea* possesses protein-polysaccharide complexes (Kim et al. 1983). In Asian countries such as China and Japan, research related to *Armillaria mellea* focuses mainly on the artificial production of fruiting bodies to be used as medicinal mushrooms (Li et al. 1993).

Kim et al. (1992) reported that the fruiting bodies were best produced in sawdust from *Quercus* spp. Shim et al. (2006) induced the formation of primordia of *Armillaria mellea* (IUM 949) in a growth chamber with illumination (350 lux) for 12 h and relative humidity of 85% for 1 day and incubated at 16 °C; using sawdust as a substrate, primordia were formed 10 d after colonization was completed and 7 days later after fruiting bodies were produced. In another study, dehydrated wild fruit bodies showed high carbohydrate content (81.25%), and less ash, fat, and protein (8.84, 1.97, and 1.81%, respectively). Mannitol was the main free sugar, while malic acid was the most abundant organic acid. The δ-tocopherol was the dominant form of tocopherols with 42.41 µg/100 g in dry weight (Kostić et al. 2017).

Ustilago maydis (DC.) Corda Huitlacoche is the name for *Ustilago maydis* that the Aztecs applied to the young and edible fruiting bodies (gall growing on the maize ears). It is a basidiomycete belonging to the *Ustilaginaceae* family and to the order *Ustilaginales* (Index Fungorum 2017). It is a causative agent of corn smut (*Zea mays* L.); fresh gills are used in Mexican cuisine. The fungus exhibits two distinct forms in its life cycle, a haploid nonpathogenic unicellular form and a filamentous pathogenic dikaryotic form. The cellular fusion between haploids gives place to the dikaryotic filament, which only can grow in the plant, where it induces the tumor formation (Valverde et al. 1995). It has been considered as an alternative crop due to the increase of its popularity as food (Venegas et al. 1995) and has been introduced into haute cuisine mainly due to its excellent taste (Ruiz-Herrera et al. 1998). But also for its nutritional content, it has approximately 11–16% protein with high content of lysine, fiber (16–23%), lipids (1.6–2.3%) among which are essential fatty acids such as linoleic and linolenic acids, ash (5–7%), and carbohydrates 55–66%) (Valverde et al. 1995).

It is an infection tolerated and sometimes even promoted by the same corn producers of the Central-South region of Mexico, since it is consumed and sold in
regional markets and canned by some food companies. Its sale represents a significant economic income, as its price far exceeds the price of corn grain (Villanueva
1997). Work has been done to implement the crop; several methods of inoculation
have been used, including sprinkling, injecting, and spraying spores; the method
that has given the best results for the production of galls is by injection of haploid
mycelium generated by fusion of opposite mating type (heterothallic) (Valverde
et al. 1995).

References

Amend A, Keeley S, Garbelotto M (2009) Forest age correlates with fine-scale spatial structure of
    Matsutake mycorrhizas. Mycol Res 113:541–551
Arnolds E (1995) Conservation and management of natural populations of edible fungi. Can J Bot
    73(Suppl. 1):S987–S998
Bao D, Gong M, Zheng H, Chen M, Zhang L, Wang H, Jian J, Wu L, Zhu Y, Zhu G, Zhou Y, Li
    C, Wang S, Zho Y, Zhao G, Zhou Y (2013) Sequencing and comparative analysis of the straw
    mushroom (*Volvariella volvacea*) genome. PLoS One 8(3):e58294
Boa E (2004) Wild edible fungi. A global overview of their use and importance to people. Non-
    Wood Forest Products 17, FAO, Rome
Bulman SR, Visnovsky SR, Hall IR, Guerin-Laguete A, Wang Y (2010) Molecular and morpholog-
    ical identification of truffle producing *Tuber* species in New Zealand. Mycol Prog 9:205–214
Cano JM, Berrocal-Lobo M, Domínguez-Núñez JA (2017) Growth of *Amanita caesarea* in the
    presence of *Pseudomonas fluorescens* and *Bacillus cereus*. Fungal Biol 121(9):825–833
Ceruti A, Fontana A, Nosenzo C (2003) Le specie europee del genere *Tuber*: una revisione storica,
    a di Regione Piemonte. Museo Regionale di Scienze Naturali, Torino
Chang ST (1977) The origin and early development of straw mushroom cultivation. Econ Bot
    31:374–376
Chang ST (1993) Mushroom biology: the impact on mushroom production and mushroom prod-
    ucts. In: Chang ST, Buswell JA, Chiu SW (eds) Mushroom biology and mushroom products.
    The Chinese University Press, Hong-Kong
Chang ST, Ling KY (1970) Nuclear behavior in the Basidiomycete, *Volvariella volvacea*. Amer J
    Bot 57:165–171
Chang ST, Miles PG (1992) Mushrooms biology a new discipline. Mycologist 6:64–65
Chang ST, Miles PG (2004) Mushrooms: cultivation, nutritional value, medicinal effect, and envi-
    ronmental impact. CRC Press, Boca Raton
Chen WA, Stamets P, Cooper RB, Huang NL, Han SH (2000) Ecology, morphology, and morpho-
    genesis in nature of edible and medicinal mushroom *Grifola frondosa* (Dicks.: Fr.) S.F. Gray-
    Maitake (Aphyllophoromycetideae). Int J Med Mushrooms 2:221–228
China Edible Fungi Association (CEFA) (2008) The yearbook 2007 of China edible fungi. China
    Mushroom Market Network, Beijing
Daza A, Manjón JL, Camacho M, De La Osa LR, Aguilar A, Santamaría C (2006) Effect of carbon
    and nitrogen sources, pH and temperature on in vitro culture of several isolates of *Amanita
    caesarea* (Scop.: Fr.) Pers. Mycorrhiza 16(2):133
Dentinger BTM, Ammirati JF, Both EE, Desjardin DE, Halling RE, Henkel TW, Moreau PA,
    Nagasawa E, Soytong K, Taylor AF, Watling R, Moncalvo JM, McLaughlin DJ (2010)
    Molecular phylogenetics of porcini mushrooms (*Boletus* section *Boletus*). Mol Phylogenet
    Evol 57:1276–1292

Dhar BL (2017) In: Zied DC, Giménez AP (eds) Mushrooms and human civilization. Edible and medicinal mushrooms: technology and applications. John Wiley & Sons, West Sussex, pp 1–4

Diyabalange T, Mulabagal V, Mills M, DeWit DL, Nair MG (2008) Health-beneficial qualities of the edible mushroom, *Agrocybe aegerita*. Food Chem 108:97–102

Duell G (2012) The President's Report [online]. National Conference of the Australian Truffle Growers Association. Available in http://www.trufflegrowers.com.au/wp-content/uploads/2012/09/2012-Presidents-Report.pdf

Duñabeitia MK, Hormilla S, Salcedo I, Peña JI (1996) Ectomycorrhizae synthesis between *Pinus radiata* and eight fungi associated with *Pinus* spp. Mycologia 88:897–908

Firenzuoli F, Gori L, Lombardo G (2008) The medicinal mushroom *Agaricus blazei* Murrill: review of literature and pharmaco-toxicological problems. Evid Based Complement Alternat Med 27:3–15

Gill WM, Guerin-Laguette A, Lapeyrie F, Suzuki K (2000) Matsutake–morphological evidence of ectomycorrhizal formation between *Tricholoma matsutake* and host roots in a pure *Pinus densiflora* forest stand. New Phytol 147:381–388

Gioacchini AM, Menotta M, Bertini L, Rossi I, Zeppa S, Zambonelli A, Piccoli G, Stocchi V (2005) Solid-phase microextraction gas chromatography/mass spectrometry: a new method for species identification of truffles. Rapid Commun Mass Spectrom 19:2365–2370

González V, Arenal F, Platas G, Esteve-Raventós F, Peláez F (2002) Molecular typing of Spanish species of *Amanita* by restriction analysis of the ITS region of the DNA. Mycol Res 106:903–910

Gyorfi J, Geosel A, Vetter J (2010) Mineral composition of different strains of edible medicinal mushroom *Agaricus subrufescens* Peck. J Med Food 13(6):1510–1514

Hall IR, Lyon AJE, Wang Y, Sinclair L (1998) Ectomycorrhizal fungi with edible fruiting bodies 2. *Boletus edulis*. Econ Bot 52(1):44–56

He X, Wang X, Fang J, Chang Y, Ning N, Guo H, Huang L, Huang X, Zhao Z (2017) Polysaccharides in *Grifola frondosa* mushroom and their health promoting properties: a review. Int J Biol Macromol 101:910–921

Index Fungorum (2017). http://www.indexfungorum.org/ (Accessed 26 Sept 2017)

Kerrigan RW (2005) *Agaricus subrufescens*, a cultivated edible and medicinal mushroom, and its synonyms. Mycologia 97(1):12–24

Kim JS, Choi EC, Kim HR, Lee CK, Lee CO, Chung KS, Shim MJ, Kim BK (1983) Studies on constituents of the higher fungi of Korea (XXXVII)-Antitumor components of *Armillaria mellea*. Kor J Mycol 11:151–157

Kim HJ, Ko MK, Yi CK, Sung JM (1992) Cultivation of *Armillaria mellea* mushrooms on a sawdust medium in polypropylene bags. Kor J Mycol 20:273–276

Kostić M, Smiljković M, Petrović J, Glamočlija J, Barros L, Ferreira IC, Ciric A, Soković M (2017) Chemical, nutritive composition and a wide range of bioactive properties of honey mushroom *Armillaria mellea* (Vahl: Fr.) Kummer. Food Funct 8:3239–3249

Leonardi M, Paolocci F, Rubini A, Simonini G, Pacioni G (2005) Assessment of inter- and intraspecific variability in the main species of *Boletus edulis* complex by ITS analysis. FEMS Microbiol Lett 243:411–416

Li JZ, Hu X, Peng Y (1993) Macrofungus flora of Hunan. Hunan Normal University Press, Changsha

Llarena-Hernández CR, Largeteau ML, Ferrer N, Regnault-Roger C, Savoie JM (2014) Optimization of the cultivation conditions for mushroom production with European wild strains of *Agaricus subrufescens* and Brazilian cultivars. Sci Food Agric 94(1):77–84

Martin F, Kohler A, Murat C, Balestrini R, Coutinho PM, Jaillon O, Montanini B, Morin E, Noel B, Percudani R, Porcel B, Rubini A, Amicucci A, Amselem J, Anthouard V, Arcioni S, Artiguenave F, Aury JM, Ballario P, Bolchi A, Brenna A, Brun A, Buée M, Cantarel B, Chevalier G, Couloux A, Da Silva C, Denoeud F, Duplessis S, Ghignone S, Hilselberger B, Iotti M, Marçais B, Mello A, Miranda M, Pacioni G, Quesneville H, Riccioni C, Ruotolo R, Splivallo R, Stocchi V, Tisserant E, Viscomi AR, Zambonelli A, Zampieri E, Henrissat B, Lebrun MH,

Paolocci F, Bonfante P, Ottonello S, Wincker P (2010) Périgord black truffle genome uncovers evolutionary origins and mechanisms of symbiosis. Nature 464(7291):1033–1038

Meinhardt F, Leslie JF (1982) Mating types of *Agrocybe aegerita*. Curr Genet 5:65–68

Mello A, Murat C, Bonfante P (2006) Truffles: much more than a prized and local fungal delicacy. FEMS Microbiol Lett 260(1):1–8

Melo de Carvalho CS, Sales-Campos C, Nogueira de Andrade MC (2010) Mushrooms of the *Pleurotus* genus: a review of cultivation techniques. Interciencia 35(3):177–182

Meotto F, Pellegrino S, Bounous G (1999) Evolution of *Amanita caesarea* (Scop.:Fr.) Pers. and *Boletus edulis* Bull.: Fr. Synthetic ectomycorrhizae on European chestnut (*Castanea sativa* Mill.) seedlings under field conditions. Acta Hortic 494:201–206

Moore D (2017) David Moore's world of fungi: where mycology starts. http://www.davidmoore.org.uk/Sec04_12.htm

Ney KH, Freytag WG (1980) Stempilz aroma (*Boletus edulis*). Gordian 80(12):304–305

Ogawa M, Umehara T, Kontani S, Yamaji K (1978) Cultivating method of the mycorrhizal fungus, *Tricholoma matsutake* (Ito et Imai) Sing. (1) Growing method of pine saplings infected with *T. matsutake* in the field. J Jap Forest Soc 60:119–128

Ohiri RC, Bassey EE (2017) Evaluation and characterization of nutritive properties of the jelly ear culinary-medicinal mushroom *Auricularia auricula-judae* (Agaricomycetes) from Nigeria. Int J Med Mushrooms 19(2):173–177

Olivier JM, Savignac JC, Sourzat P (1996) Truffe et trufficulture. Fanlac, Perigueux, France.

Pandey A, Soccol CR, Nigam P, Brand D, Mohan R, Roussos S (2000) Biotechnological potential of coffee pulp and coffee husk for bioprocesses. Biochem Eng J 6:153–162

Park WH, Lee HD (1991) Wild fungi of Korea in color. Kyo-Hak Publishing Co, Seoul

Park WH, Lee HD (2005) Wild fungi of Korea. Kyo-Hak Publishing Co. Ltd

Poitou N, Mamoun M, Delmas J (1982) Quelques résultats obtenus concernant la mycorhization de plantes-hotes par les champignons mycorhiziens comestibles. Les Mycorhizes: biologie et utilization Les Colloques de l'INRA 13:295–301

Psurtseva N (2005) Modern Taxonomy and Medicinal Value of the *Flammulina* Mushrooms. Int J Med Mushrooms 7:449

Rodrigues DM, Freitas AC, Rocha-Santos TA, Vasconcelos MW, Roriz M, Rodríguez-Alcalá LM, Gomes AMP, Duarte AC (2015) Chemical composition and nutritive value of *Pleurotus citrino-pileatus* var. *cornucopiae*, *P. eryngii*, *P. salmoneo stramineus*, *Pholiota nameko* and *Hericium erinaceus*. J Food Sci Technol 52(11):6927–6939

Royse DJ (1995) Specialty mushrooms: cultivation on synthetic substrate in the USA and Japan. Interdisciplin Sci Rev 20:1–10

Ruiz-Herrera J, Martínez-Espinoza AD (1998) The fungus *Ustilago maydis*, from the aztec cuisine to the research laboratory. Int Microbiol 1:149–158

Sakamoto Y, Ando A, Tamai Y, Yajima T (2007) Pileus differentiation and pileus-specific protein expression in *Flammulina velutipes*. Fungal Genet Biol 44:14–24

Salerni E, Perini C (2004) Experimental study for increasing productivity of *Boletus edulis* sl in Italy. For Ecol Manage 201:161–170

Sánchez-Ramírez S, Tulloss RE, Amalfi M, Moncalvo JM (2015) Palaeotropical origins, boreo-tropical distribution and increased rates of diversification in a clade of edible ectomycorrhizal mushrooms (*Amanita* section Caesareae). J Biogeogr 42(2):351–363

Savoie JM, Largeteau ML (2011) Production of edible mushrooms in forests: trends in development of a mycosilviculture. Appl Microbiol Biotechnol 89(4):971–979

Shim JO, Chang KC, Lee YS, Park CH, Kim HY, Lee U, Lee T, Lee MW (2006) The fruiting body formation of *Armillaria mellea* on oak sawdust medium covered with ground raw carrots. Mycobiology 34(4):206–208

Sitta N, Davoli P (2012) Edible ectomycorrhizal mushrooms: international markets and regulations. In: Zambonelli A, Bonito GM (eds) Edible ectomycorrhizal mushrooms. Springer, Berlin Heidelberg, pp 355–380

Stamets P (2000) Growing gourmet and medicinal mushrooms. Ten Speed Press, Berkeley

Stamets P (2005) Notes on nutritional properties of culinary-medicinal mushrooms. Int J Med Mushrooms 7:109–116

Stojković D, Reis FS, Barros L, Glamočlija J, Ćirić A, Van Griensven LJ, Soković M, Ferreira IC (2013) Nutrients and non-nutrients composition and bioactivity of wild and cultivated *Coprinus comatus* (OF Müll.) Pers. Food Chem Toxicol 59:289–296

Tang L, Xiao Y, Li L, Guo Q, Bian Y (2010) Analysis of genetic diversity among Chinese *Auricularia auricula* cultivars using combined ISSR and SRAP markers. Curr Microbiol 61(2):132–140

Téllez-Téllez M, Díaz-Godínez G (2017) *Pleurotus ostreatus* as a nutraceutical food. In: Mervyn P, Gwynn I (eds) *Pleurotus* mushrooms: ecology, cultivation and uses. Nova Science Publisher, New York, pp 57–91

Thongbai B, Rapior S, Hyde KD, Wittstein K, Stadler M (2015) *Hericium erinaceus*, an amazing medicinal mushroom. Mycol Prog 14(10):91

Tominaga Y (1978) *Tricholoma matsutake*. In: Chang ST, Hayes WA (eds) The biology and cultivation of edible mushrooms. Academic Press, London, pp 683–697

Tsai SY, Tsai HL, Mau JL (2007) Antioxidant properties of *Agaricus blazei, Agrocybe cylindracea*, and *Boletus edulis*. LWT-Food Sci Technol 40(8):1392–1402

Ullrich R, Nüske J, Scheibner K, Spantzel J, Hofrichter M (2004) Novel haloperoxidase from the Agaric Basidiomycete *Agrocybe aegerita*. Appl Environ Microbiol 70:4575–4581

Vaario LM, Guerin-Laguette A, Gill WM, Lapeyrie F, Suzuki K (2000) Only two weeks are required for *Tricholoma matsutake* to differentiate ectomycorrhizal Hartig net structures in roots of *Pinus densiflora* seedlings cultivated on artificial substrate. Can J For Res 5(4):293–297

Valverde ME, Paredes-López O, Pataky JK, Guevara-Lara F, Pineda TS (1995) Huitlacoche (*Ustilago maydis*) as a food source-biology, composition, and production. Crit Rev Food Sci Nutr 35(3):191–229

Van Griensven LJLD (1988) The cultivation of mushrooms. Darlington Mushroom Laboratories, Somycel

Venegas PE, Valverde ME, Paredes-López O, Pataky JK (1995) Production of the edible fungus huitlacoche (*Ustilago maydis*): effect of maize genotype on chemical composition. J Biosci Bioeng 80(1):104–106

Villanueva VC (1997) Huitlacoche (*Ustilago maydis*) as a food in Mexico. Micol Neotrop Apl 10:73–81

Volman JJ, Helsper JPFG, Wei S, Baars JJP, Van Griensven LJLD, Sonnenberg ASM, Mensink RP, Plat J (2010) Effect of mushroom-derived β-glucan-rich polysaccharide extracts on nitric oxide production by bone marrow-derived macrophages and nuclear factor-kB transactivation in Caco-2 receptor cells: can effects be explained by structure? Mol Nutr Food Res 54:268–276

Wang X (2012) In: Zambonelli A, Bonito GM (eds) Edible ectomycorrhizal mushrooms: Current knowledge and future prospects. Springer, Berlin

Yamada A, Kanekawa S, Ohmasa M (1999) Ectomycorrhizal formation of *Tricholoma matsutake* on *Pinus densiflora*. Mycoscience 40:193–198

Yamanaka K, Namba K, Tajiri A (2000) Fruit body formation of *Boletus reticulatus* in pure culture. Mycoscience 41:189–191

Yan PS, Luo XC, Zhou Q (2004) RAPD molecular differentiation of the cultivated strains of the jelly mushrooms, *Auricularia auricula* and *A. polytricha*. World J Microbiol Biotechnol 20:795–799

Yang GL, Xue HB (2000) Specialized cultivation manual about edible and medicinal mushroom. China Agricultural Press. Beijing, China, pp 361–368

Yun W, Hall IR (2004) Edible ectomycorrhizal mushrooms: challenges and achievements. Can J Bot 82(8):1063–1073

Chapter 10
Recent Developments in Shiitake Mushrooms and Their Nutraceutical Importance

Richa Salwan, Shabnam Katoch, and Vivek Sharma

10.1 Introduction

The mushroom production across the world has been categorized between Europe, North America, and Australia on one side and Southeast Asia such as China and Japan on the other side. European and American production is almost exclusively linked to white button mushroom (*Agaricus bisporus*), whereas in Asian counterpart, it is dominated by shiitake, paddy straw mushrooms (*Volvariella volvacea*) and the oyster mushrooms (*Pleurotus* spp.) (Chang and Miles 1991; Whiteford and Thurston 2000). Shiitake (*Lentinula*) mushroom belongs to division *Basidiomycotina*, which are wood-decaying fungi. *Lentinula* are gregarious in nature and especially reported on fallen wood of broadleaf deciduous trees, especially on members of family Fagaceae (Tokimoto and Komatsu 1978; Przybylowicz and Donoghue 1988) such as chestnut, maple, sweet gum, oak, beech, poplar, alder, hornbeam, and mulberry under warm and moist climatic conditions (Wasser and Weis 1997). The geographic distribution of shiitake mushroom extends under natural conditions from different countries such as India, Japan, china, New Zealand, Bhutan, and Nepal (Pegler 1983; Shimomura et al. 1992). Shiitake were also reported growing on conifer wood in Kazakhstan representing Asia, South America, and North America, Australia, and other continents (Samgina and Agailkovie 1981). However, it is reported especially on Fagaceae; therefore, its identity from Kazakhstan may have

R. Salwan
College of Horticulture and Forestry, Dr YS Parmar University of Horticulture & Forestry, Hamirpur, HP, India

S. Katoch
College of Agriculture, Chandigarh University, Mohali, Punjab, India

V. Sharma (✉)
University Centre for Research and Development, Chandigarh University, Mohali, Punjab, India

© The Author(s), under exclusive license to Springer Nature Switzerland AG 2021
X. Dai et al. (eds.), *Fungi in Sustainable Food Production*, Fungal Biology, https://doi.org/10.1007/978-3-030-64406-2_10

been misidentified. Only a single collection of *Lentinula guarapiensis* is reported from Paraguay till 1983 (Pegler 1983). The reports on shiitake distribution from Papua New Guinea region and the Pacific islands are poorly reported (Bisen et al. 2010).

Shiitake are one of the largest cultivated mushrooms worldwide due to their importance in food and therapeutic applications. It was first macrofungus to attain the focus in modern biotechnology. In food, it is also used as appetizers as well as nourishing (Xu et al. 2014). It contributes to about 17% of the total global consumption of mushrooms. In Japan and the Far East, shiitake is one of the most edible mushrooms (Hiraide et al. 2004; Lee et al. 2017). In therapeutics, shiitake mushrooms are documented for treating ailments related to immune suppression, cancer and allergies treatments against viral and fungal infections, bronchial inflammation, heart diseases, treating blood cholesterol, diabetes, and several other diseases. Shiitake or golden oak mushroom is explored for scale cultivation and production. Its production is responsible for the economy of northeastern region of India. Shiitake mushrooms are of paramount importance due to their unique taste and health and nutraceutical benefits. The pharmacological properties of shiitake mushrooms are attributed to polysaccharides such as lentinan, eritadenine, shiitake mushroom mycelium, and culture media extracts. It is composed of 58–60% of carbohydrates, 20–23% of proteins, 9–10% of fibers, 3–4% of lipids, and 4–5% of ash (Wasser 2003; Xu et al. 2014).

The delimitation of *Lentinula* species is controversial (Pegler 1983; Hibbett 1992; Hibbett et al. 1995). It was initially represented by three species such as *L. edodes* from continental and northeast Asia and *L. lateritia* from tropical Asia and Australia and *L. novaezelandieae* of New Zealand (Pegler 1983). However, based on mating compatibility, all of these species were found interfertile and hence categorized as a single species (Shimomura et al. 1992). Thereafter, based on phylogenetic analysis, four distinct lineages (groups I–IV) of shiitake mushroom in Asia–Australia were reported (Hibbett et al. 1995). Group I represented Japan, Borneo, and Thailand, whereas group II represented Tasmania and Papua New Guinea. Group III was restricted to Papua New Guinea. However, *Lentinula* from several regions such as China and Australia were not included (Hibbett et al. 1998). The phylogenetic analysis based on ITS sequences revealed five major lineages of *Lentinula* (groups I–V). Based on bootstrap values of 80–100%, these groups were categorized as monophyletic. However, later on root ambiguities (bootstrap value of 70%) proved that only groups II, III, and V were monophyletic (Hibbett et al. 1998). The major establishment of this study approved a fifth lineage (group V) of shiitake mushrooms in continental Asia which includes eastern China and Nepal. Group V from Himalayan regions representing Nepal contains only one isolate (Hibbett et al. 1998).

In India, several *Lentinus* species have been reported along with *Panus* from the Western Ghats (Pegler 1983a, b; Manimohan et al. 2004; Natarajan et al. 2005; Farook et al. 2013; Senthilarasu 2014). Different species of *Lentinus* are known as *L. bambusinus* (Kumar and Manimohan 2005), *L. polychrous* (Manimohan et al. 2004; Mohanan 2011), *L. patulus* (Mohanan 2011), *L. sajor-caju* (Manimohan et al.

2004; Pradeep and Vrinda 2007; Varghese et al. 2010; Mohanan 2011), *L. dichola-mellatus*, *L. tigrinus* (Mohanan 2011), and *L. squarrosulus* (Sharma et al. 1985; Florence and Yesodharan 2000; Manimohan et al. 2004; Pradeep and Vrinda 2007; Varghese et al. 2010; Mohanan 2011; Senthilarasu 2014). On the other side, the phylogenetic characterization of wild edible mushroom of Nagaland and India identified *L. edodes*, *L. squarrosulus*, *L. tigrinus*, *L. sajor-caju*, and *L. squarrosulus* (Ao et al. 2019). Recently, a group of 44 simple sequence repeat (SSR) markers has been developed for shiitake mushroom cultivars based on the complete genomic information of *L. edodes*. These markers can be explored for molecular breeding, determination of cultivars, and other applications of shiitake mushrooms (Lee et al. 2017).

Here in this chapter, we have discussed the taxonomic view, cultivation practices of mushroom, and its importance in health and nutraceuticals sectors of shiitake which can play a big role in increasing the livelihood of farmers in India.

10.2 Historical Biogeography of Shiitake Cultivation

Lentinula mushrooms represent a group of wood-decaying higher fungi. The genus is known for shiitake mushroom cultivation. Lentinula has been reported in the wild from continents such as Australia, Asia, and America. *Lentinula boryana* has been reported from Northern to South America and the Gulf Coast of North America, whereas a single collection of *L. guarapiensis* has been reported from Paraguay (Pegler 1983). Species demarcation of shiitake is controversial. As described earlier, shiitake was initially classified into three species: *L. edodes* from continental and northeast Asia, *L. lateritia* from tropical Asia and Australia, and *L. novaezelandieae* from New Zealand (Pegler 1983). However, since all these species were interfertile, therefore it is revealed that all of them should be considered a single species (Shimomura et al. 1992). Based on phylogenetic analysis, four distinct lineages (I–IV) of *Lentinula* were reported in Asia–Australia, and it was proposed that these lines could be categorized as phylogenetic species (Hibbett et al. 1995). The group I belonged to Thailand, Japan, and Borneo, whereas the group II was reported from Papua New Guinea and Tasmania. The III group was restricted to New Zealand, whereas group IV was restricted to Papua New Guinea. However, many *Lentinula* populations from China and Australia not represented in previous study were included (Hibbett et al. 1998). In another study, isolates of *Lentinula* from the Old and New World were found incapable of mating with each other (Guzman et al. 1997). Two species of *L. sensu* Pegler, i.e., *L. boryana* (Berk. & Mont.) Pegler and *L. guarapiensis* (Speg.), were reported in the New World. The phylogenetic analysis based on rDNA revealed that the members of *Lentinula* of both the New World and Old World were classified as monophyletic. The phylogenetic analysis based on ITS analysis revealed seven unique *Lentinula* species (Hibbett et al. 1998; Hibbett and Avenue 2001).

10.3   Life Cycle and Methods of Shiitake Cultivation and Increasing Its Self-Life

The shiitake cultivation includes mycelial growth, formation of brown film, initiation of primordium, and fruiting body formation (Tang et al. 2013). The formation of brown film formation is essential and unique for shiitake fruiting production. The brown film formation includes the accumulation of pigments on older mycelial surface. The biofilm also maintains the water content in the growing medium, UV radiation absorption, and protection to the inner mycelia and helps in higher production as well as quality (Belozerskaya et al. 2017; Yoo et al. 2019; Yan et al. 2020). During this entire process, electron-dense material accumulates between the plasma membrane and cell wall space and steadily also accumulates on the outside of the cell wall which leads to cell wall stratification. The electron-dense material accumulated on the cell wall is similar to the primordium mycelial network (Butler and Day 1998; Vetchinkina et al. 2017). However, the variations on the deposition of electron-dense materials in sawdust-containing medium were reported to be higher than PDA and in mycelial tissues such as primordium or surface mycelium. The electron microscopic-based studies revealed that melanin precursor is transported in lipid vesicles and cytoplasmic vacuoles to the outside of cell wall (Butler and Day 1998; Vetchinkina et al. 2017; Yan et al. 2020).

Shiitake belongs to division *Basidiomycota*, class *Basidiomycetes*, order *Agaricales*, and family *Agaricaceae* (Wu et al. 2010). The genus is known to produce sexual basidiospores. Fruiting bodies known as basidiocarps contain the basidiospore-producing structure called basidium (Terashima and Matsumoto 2004). Individual basidium contains four basidiospores each containing one haploid nucleus. Basidiospores once released and get favorable conditions germinate, and the hypha contains a single nucleus. The mycelium from a single spore cannot form a fruiting body. It must pair with hypha from another spore of the opposite mating. Hyphal tips fuse with each other to produce a small, compact dense ball-like compartment known as primordium (Bisen et al. 2010). It is small in size (0.5–1 mm) and often buried as loose fuzzy mycelium in soil which is difficult to locate. However, it contains the raw material required for the fruiting body formation. Once the conducive conditions such as rainfall and appropriate temperature prevail, the mycelial network collectively pumps nutrients and water into the primordium for its rapid expansion which persists for a period of 12–96 h (Leatham 1982). After mating, the characteristic "fairy rings" are formed which give rise to mushrooms due to prior exploration of nutrients over a large area. The formation of fruiting bodies occurs in the older hyphae where nutrients are exhausted and there is scarcity of food.

Lentinula edodes, known as wood-degrading fungi, is traditionally grown on logs and sawdust-based materials. A combination of sawdust, wood wastes, and straw or rice bran is also used for a large-scale production (Ohga and Kitamoto 1997: Yamanaka 1997). Under lab conditions, the pure culture of shiitake is maintained on potato dextrose agar (PDA) medium. However, economical and alternate medium for pure brood stocks of shiitake rice wastewater has been used. The composition of rice wastewater is known to be similar to PDA. The basic culturing of shiitake mycelium requires carbon sources such as glucose, sucrose, and lignocellulose along with ammonium chloride to grow (Hoa and Wang 2015; Setiati et al. 2019). The addition of Cu, Zn, or Se to produce media enhanced the anti-inflammatory potential of *L. edodes* mycelial extract, thus indicating its potential as a natural anti-inflammatory dietary supplement (Muszyńska et al. 2019). The optimum growth temperature of spawn is 25–28 °C, 80–85% humidity, diffused light, and ventilation for 80–90 days. Shiitake can be on a sterilized mixture of sawdust and wood chips. The total cultivation time is of 110–120 days. The shelf-life of mushroom is 2–3 days at 25–30 °C and 10–12 days at 4 °C. In India, the natural growth includes temperature between 20 and 25 °C. All year-round cultivation can be done in Chikmagalur; Coonoor; Ooty; Kodagu; Kodaikanal; northeastern states of Manipur, Mizoram, Meghalaya, and Nagaland; and part of Himachal Pradesh and Arunachal Pradesh, whereas seasonal cultivation is performed in other parts of the country. During harvesting, fruiting bodies are brown in color and stalks are white. It is marketed either as fresh or dry or as mushroom powder, whereas the residual mushroom-growing substrate is an excellent organic manure.

Since the mushrooms are having a limited shelf-life depending upon the storage conditions, then their nutritional quality starts deteriorating (Jiang et al. 2014). Enhancing shelf-life of shiitake can be valuable due to high demands in several countries (Gavahian et al. 2019a, b). Efforts have been made from time to time to increase its shelf-life. For example, use of plasma gas and plasma-activated water has been found an attractive method for deactivating microbial spoilage without affecting the physical sensory and nutritional aspects of the mushroom (Gavahian et al. 2019a, b; Pankaj et al. 2017). Drying using hot air or microwave or vacuum and a combination of microwave vacuum have been tested. All of the methods led to a significant increase in total free amino acid contents and sulfur compounds. Hot air and vacuum drying was found to increase vitamin B12 in dried mushrooms, whereas microwave vacuum-assisted drying maintained high amounts of active amino acids related to taste, enhanced nutrient retention, as well as color (Tian et al. 2016). The microwave and a combination of microwave vacuum drying resulted in the highest and lowest amount of volatile compounds, respectively. Further, morphological examinations such as maintenance of uniform honeycomb network, quality of dried product, and volatile compounds and microwave vacuum drying methods have been tested (Tian et al. 2016). The other methods of active constituent's extraction such as ultrasound-assisted, pulsed electric field, supercritical fluid, and ultrahigh pressure-assisted methods have been used. These methods are energy efficient and environmental friendly (Sun et al. 2019). The preservative effects of plasma-activated water for enhancing the shelf-life of *Agaricus bisporus* have been tested (Xu et al. 2016). The cold plasma is also considered as a potential alternate in comparison with conventional preservation-based techniques (Gavahian et al. 2018; Potluri et al. 2018). It is mainly due to the production of chemically active molecules such as reactive oxygen species (ROS) in the plasma gas and PAW-based methods (Gavahian et al. 2018; Gavahian and Cullen 2019; Gavahian and Khaneghah 2019).

10.4 Metabolic Profiling

The smell of mushroom is one of the main criteria in determining its quality. The fresh shiitake mushroom releases only a slight odor. However, gradually a characteristic sulfurous aroma is released upon drying and crushing (Yasumoto et al. 1976). The ingredients of edible mushrooms can be influenced by several factors including type of species, part of and maturity stage of mushroom, methods adopted for processing, and storage conditions (Gao et al. 2020). In several studies, the volatile metabolites of dried shiitake mushroom have been identified. The 1,2,3,5,6-pentathiepane, also known as "lanthionine," is reported as a key compound (Hiraide et al. 2004). The other aroma components such as alcohols, ketones, sulfides, alkanes, and fatty acids have been reported. The presence of calciferol (vitamin D2) and vitamin-B complex such as B1 (thiamine), B2 (riboflavin), and B12 (niacin) along with fatty acids and amino acids has been observed. Besides the characteristic aroma of shiitake mushrooms due to the presence of 1,2,3,5,6-pentathiepane, the other volatile flavor compounds such as matsutakeol (octen-1-ol-3) and ethyl-n-amyl ketone are also reported (Bisen et al. 2010). The nutritious compounds such as 5'-guanylic acid and vitamin D2 are also reported in higher amount in dried shiitake mushroom. Besides this, several other sulfur-containing metabolites and alcohols are also reported. Researchers have also reported 1,2,4-trithiolane and 1,2,4,6-tetrathiepane in dried shiitake mushrooms; the former was identified as indicator of the smell. Lenthionine ($C_2H_4S_5$) was initially identified in the late 1960s (Morita and Kobayashi 1967; Wada et al. 1967a, b), and it was found to decompose during analysis (Charpentier et al. 1986). Thereafter, other sulfur-containing compounds such as 1,2,4-trithiolane, 1,2,4,6-tetrathiepane, and 1,2,3,4,5,6-hexathiepane were also identified (Morita and Kobayashi 1967). Other compounds such as dimethyl disulfide, dimethyl trisulfide, 1,2,4-trithiolane, and one sulfur hereafter S-containing compound with molecular formula $C_2H_4S_4$ are also identified. 1,2,4-Trithiolane was the only cyclic S compound identified in homogenized fresh mushroom (Kameoka and Higuchi 1976; Chen et al. 1984).

The origin of cyclic S compounds from lentinic acid, derived from γ-glutamyl cysteine sulfoxide, in shiitake mushroom dated back to late 1970 (Iwami et al. 1975a, b; Yasumoto et al. 1976). The γ-glutamyl transpeptidase and cysteine sulfoxide lyase were found to convert lentinic acid into cyclic compounds containing S. Later in 1978, it was found in dried mushroom; the pH and temperature affect the lenthionine biosynthesis (Ito et al. 1978) during rehydration (Chen and Ho 1986b). Besides lenthionine, the compounds such as 1,2,4,6-tetrathiepane and 1,2,3,4,5,6-hexathiepane (Morita and Kobayashi 1967, Chen and Ho 1986a, b) and 8-carbon alcohols responsible for the smell of fresh shiitake (Chen and Ho 1986b; Cronine and Ward 1971; Kameoka and Higuchi 1976; Ito et al. 1978) were not reported in dried shiitake mushroom. The compounds responsible for sensory intensity were found to be linked to S-containing compounds by H_2S, SO_2, and other compounds. The presence of trace amount of methyl β-methyl-thiopropionate and p-menthen-8-thiol is also found responsible for pineapple and grapefruit types of flavors, respectively. However, the role of sulfur-containing compounds in imparting smell to dried shiitake mushroom is not established (Nishikawa et al. 1967; Wada et al. 1967a, b; Yasumoto et al. 1971; Hiraide et al. 2004).

The carbohydrates and glycoproteins of shiitake mushroom such as lentinan and LEM and LAP are very good source of nutrition. The lentinans are polysaccharides $[(C_6H_{10}O_5)n]$ of high molecular weight 5×10^5 Da present in the cell wall of fruiting bodies. It is soluble in water and free from phosphorus and sulfur. It is present as a triple helix structure which is composed of glucose molecules linked to branched backbone via β-(1-3)-linkages and β-(1-6)-glucose side chain (Wasser and Weis 1997; Wasser 2002). The structural orientation of glucose moieties in a helix is known to play important biological and pharmacological roles (Mizuno 1999). The other biological molecules known as glycoproteins such as *L. edodes* mycelia (LEM) and LAP have been obtained from mushroom mycelia and culture media, respectively. The LEM contains nucleic acid derivatives, vitamin B and ergosterol compounds and is found to have an antitumor activity. The immune-active molecule EP3 has been obtained from LEM fractionation (Suzuki et al. 1990). The other peptide–polysaccharide-based complex known as KS-2-α-mannan peptide KS-2 has been obtained from the extraction of *L. edodes* mushroom upon treatment with hot water and precipitation with ethanol. It is found effective against Sarcoma 180 and Ehrlich's carcinoma (Fujii et al. 1978). The other molecule known as eritadenine, a secondary metabolite isolated from *L. edodes* mycelia (400–700 mg/kg of dry matter), is identified for its role in lowering cholesterol level (Chibata et al. 1969). It was formerly assigned as lentinacin (Chibata et al. 1969) and lentysine (Kamiya et al. 1969; Rokujo et al. 1970; Saito et al. 1975). Additionally, sugars such as trehalose, arabitol, glycerol, mannose, mannitol, and arabinose have been reported. The water-soluble material dietary fibers composed of β-glucan and protein along with heteropolysaccharides, lignin and chitin, are also reported in the cell wall (Bisen et al. 2010).

10.5 Genome Organization and Gene Mining for Unique Aroma of Shiitake and Shelf-Life Enhancement

The genome of monokaryotic *L. edodes* sequenced on Illumina HiSeq 2000 and 2500 platform is 41.8–46.1 Mb in size. The complete genome encodes 13,028–14,889 predicted genes (Chen et al. 2016; Shim et al. 2016). Of these, only 11.5% of the depicted proteins in *L. edodes* are found to share orthologs with other fungal species, whereas about 22% of the proteins were unique to *L. edodes* (Chen et al. 2016; Shim et al. 2016). The enzymes γ-glutamyl transpeptidase and C-S lyase are involved in the formation of lenthionine; a main volatile flavor is produced from lentinic acid (a γ-L-glutamyl-cysteine sulfoxide precursor) (Chen et al. 2016). The enzymes γ-glutamyl transpeptidase and C-S lyase are involved in the formation of lenthionine. A total of 7 *ggt* and 5 Csl encoding genes have been reported in the genome of *L. edodes* (Chen et al. 2016). Across 26 fungi, in general 2–4 *ggt* genes have been identified. Comparatively, high number of *ggt* genes revealed better ability of *L. edodes* in catalyzing lentinic acid to L-cysteine sulfoxide derivative. The C-S lyase genes were found to be a novel cysteine desulfurase (EC 2.8.1.7) with cysteine sulfoxide lyase activity (Chen et al. 2016).

In another study, de novo assembly performed using paired-end short reads and long reads and transcriptome sequencing (RNA-seq) predicted 12,959 genes from the life cycle of *L. edodes*. It was also found that shiitake lacks genes for lignin peroxidase. Further, genes involved in the postharvest fruiting body degradation of fruiting bodies using transcriptome and serial analysis of gene expression (SAGE) analysis identified upregulation of cell wall hydrolysis-related genes such as glycosyl hydrolase (GH) families GH5, GH16, GH30, GH55, and GH128; thaumatin-like proteins; chitinases of GH18 and GH20; as well as putative chitosanase of GH75, during the autolysis of fruiting body. Besides this, several putative transcription factors including high-mobility-group domains, zinc finger domains, and cell death-related proteins were found upregulated during postharvest. Therefore, genes identified for their role in fruiting body autolysis can be targeted for the breeding-improved strains with better fresh life (Sakamoto et al. 2017).

10.6   Nutraceutical Importance

The shiitake mushrooms are consumed as foods and medicine in Asia over 2000 years. The edible mushrooms are rich source of polysaccharides, proteins, terpenoids, polyphenols, vitamins, and dietary fibers (Moradali et al. 2007; Asaduzzaman and Mousumi 2012; Reis et al. 2012; Xu et al. 2014). The polysaccharides, proteins, saponins, triterpenes, and flavonoids of edible mushrooms are a part of healthy foods (Gu et al. 2016). Worldwide, shiitake (*L. edodes*) is the second largest cultivated edible mushroom and accounts for approximately 25% of total edible mushroom production (Jiang 2011). The commercial production of Shiitake (*L. edodes*) is available in the USA, Brazil, India, Australia, and Europe (Mason et al. 2004). The production of shiitake mushroom is over 1,30,000 tons per year, and 45% of produce is sold fresh. The remaining dried preparation of *L. edodes* is available in different forms. The high nutritional value is attributed to the presence of several bioactive compounds such as lentinan, vitamins (B1, B2, and C), dietary fibers, ergosterol, niacin, and minerals (Choi et al. 2006). Presently, the shiitake mushrooms are also consumed in western countries (Liu et al. 2013). Fruit bodies of *L. edodes* are composed of 88–92% water, and remaining contents include proteins, lipids, carbohydrates, as well as vitamins and minerals. The mushrooms are a source of vitamins, provitamin D2 (ergosterol), which under ultraviolet (UV) light and heat is converted to calciferol (vitamin D2). It also contains vitamin-B complex including B1 (thiamine), B2 (riboflavin), and B12 (niacin) (Dermiki et al. 2013; Chen et al. 2015; Poojary et al. 2017). The vitamin such as D although not reported is routinely used foodstuffs, but edible mushrooms are one of the greatest source of its precursor ergosterol (Mattila et al. 2000). Upon photolysis, ergosterol is converted to previtamin D2. Subsequent UV treatment leads to the isomerization of previtamin D2 to vitamin D2 (Keegan et al. 2013). A combination of fine grinding and UV-C treatment increases the conversion of ergosterol to vitamin D2 synergistically. The fatty acids constitute up to 3.38% of the total lipids along with amino acids. The presence and variation of monosodium glutamate, nucleotides, free amino acids, small-sized molecular weight peptides, organic acids, and sugars are attributed to delicious flavor Mizuno. Further, organic acids such as acetic acid, fumaric acid, oxalic acid, malic acid, α-ketoglutaric acid, lactic acid, formic acid, and glycolic acid also contribute to its flavor (Bisen et al. 2010). Additionally, the combined treatment enhanced the water holding capacity, water solubility, polysaccharide dissolution, and increase in antioxidant profiles due to accumulation of tannin contents in mushrooms without affecting the physiochemical properties (Xu et al. 2019).

The rheological quality of wheat starch supplemented with *Lentinus edodes* β-glucan (LEBG) revealed that the amount of slow digestible and resistant starch in mixture increases. The presence of slow digestible starch and resistant starch with addition of 20% LEBG was found to be 18.23 and 54.95%, respectively which is higher compared to the control. The predicted glycemic index reduced with the increasing concentration of LEBG. Therefore, the LEBG concentration significantly impacted the rheological property as well as in vitro digestion of starch. Moreover, the replacement of starch with 20% LEBG reduced predicted glycemic index (GI) values. Hence, starch supplemented with LEBG can be explored for low glycemic index foods and even was a good source of low GI noodle bread (Zhuang et al. 2017). The preparation of shiitake mushrooms can be ingested as a sugar-coated tablet or capsules, powder, concentrate, and syrup and/or as a medicine (Zhuang et al. 2017).

10.7 Pharmacological and Immunomodulatory Roles of *L. edodes*

Shiitake mushrooms are well explored for their various therapeutic applications. The pharmacological properties of shiitake mushrooms have been studied from a long time back dated to Ming dynasty (1369–1644). In Japanese empire, shiitake is considered as the "elixir of the life" for enhancing the energy and vigor, as anticarcinogenic, antimicrobial agents and antidiabetic, hypolipidemic activities. The fruiting bodies of edible mushrooms *Lentinula edodes* along with other edible mushrooms such as *Agaricus bisporus*, *Cantharellus cibarius*, and *Imleria badia* are a source of cholesterol-lowering drug such as lovastatin (Kała et al. 2020). The anticancer activities of *L. edodes* are contributed due to their immune-stimulating, anti-inflammatory, and antioxidant roles (Bisen et al. 2010).

Lentinus edodes has gained major attention for its immunomodulatory roles due to lentinan and other immune-boosting polysaccharides and edible hybrid mushrooms (Bhunia et al. 2010). Both water-soluble and water-insoluble types of β-glucans (Yu et al. 2010) have been isolated from *L. edodes* (Maity et al. 2013). The polysaccharides belonging to the group of β-D-glucans such as lentinan of *L. edodes* are found to boost the immune system. These polysaccharides do not cause any harm and, therefore, are considered as biological response modifier. The mechanisms responsible for antitumor effect include oncogenesis prevention, boosted immune response, and promotion of apoptosis due to direct antitumor activity (Zhang et al. 2007). In order to exhibit its biological activity, lentinan needs functional T-cell component. The lentinan and immune boosting polysacharides increases T helper (Th) cell and macrophage profileration and also stimulate the non-immunological acute phase proteins and colony-stimulating factor (CSF), which affects the macrophages, lymphocytes cells proliferation and the activation of the complement system (Bohn and BeMiller 1995; Xu et al. 2014). A water-soluble polysaccharide obtained after sepharose gel filtration of hybrid mushroom strain (obtained after fusion of protoplast of *P. florida* and *L. edodes*) resulted in activation of macrophage, thymocyte, and splenocytes (Maity et al. 2013).

The antibiotic substances such as protein-based lentin, exobiopolymer-containing sulfur-based lenthionine, purine-derived lentinosin, and lentinamycins A and B obtained from fruiting bodies of *L. edodes* have been identified. The atherosclerotic molecules such as eritadenine and lovastatin have been obtained from fruiting bodies of *L. edodes*. The shiitake mushrooms are also used for diseases, environmental allergies, bronchial inflammation, frequent colds, and flu and regulating urinary incontinence (Bisen et al. 2010). The terpenoids and sterol compounds of mushrooms are reported for antimicrobial and antioxidant activities (Zhang et al. 2013). Besides the application of shiitake mushrooms in food and medicinal industries, the carbon composite with $NiCo_2O_4$ has been developed for efficient supercapacitor applications where mushroom has been as the carbon. The high surface area of mushroom is an advantage over other carbon materials. The mushroom-C/$NiCo_2O_4$ composite is known to provide a better capacitance compared with carbon and other metal-based oxide composite electrodes (Raman et al. 2019).

10.8 Concern

The formation of formaldehyde endogenously in the shiitake mushroom is a major concern. The formation of formaldehyde was observed during analysis of flavor compounds (Okada et al. 1972). The catalysis of lentinic acid γ-glutamyl transpeptidase (GGT) and L-cysteine sulphoxide lyase (C-S lyase) was found to produce the characteristic flavor compounds and formaldehyde (Yamazaki et al. 1980; Yasumoto et al. 1976). However, the exact mechanisms of formaldehyde formation in *L. edodes* are not known. Formaldehyde is present in natural foods as well as used in dry and frozen processed food to keep it fresh. It is toxic to the eyes and can cause blindness and other respiratory problems. It is a mutagen and potentially carcinogenic (Weng et al. 2009). As per the US Environmental Protection Agency, the recommended daily dose of formaldehyde is 0.2 mg/kg of body weight. However, formaldehyde level of up to 119–494 mg/kg has been reported in dried fruiting bodies (Liu et al. 2005). In other study, formaldehyde content of UK and Chinese shiitake mushroom was up to 100–300 mg/kg (Mason et al. 2004; Liu et al. 2013). The other side effects include development of flagellate types of rashes and even severe cutaneous reaction. The dermatitis is caused by a toxic reaction to a thermolabile polysaccharide known as lentinan, a component of raw and undercooked shiitake mushrooms. The symptom appears within 2 hours or even 5 days of ingestion and can take up to 28 days to recover.

10.9   Conclusion

The potential of shiitake mushrooms in therapeutics, applied biotechnology, and various food industries is unquestionable. In medical field, due to a plethora of bio-active metabolites, it is used as a prolific source of vitamins and drugs to combat various diseases such as diabetes, cancer, and various physiological disorders. Moreover, several undemonstrated therapeutic roles of shiitake mushroom need further advanced studies. However, lack of standard protocols, marketing strategies, quality assessment, and bioactivity of compounds derived from it in human context needs to be verified. The effects of different carbon sources such as rice straw or sawdust on chemical composition and volatile and nonvolatile taste components of shiitake mushrooms are poorly understood. In parallel, precise strain identification, the mining of genome, and identification of key genes related to aroma and enhancing shelf-life of fruiting bodies by the genetic engineering of potential strains also need researcher's attentions.

References

Ao T, Ranjan C, Satyawada D, Rao R (2019) Molecular strategies for identification and characterization of some wild edible mushrooms of Nagaland, India. Molecular Biology Reports, (0123456789). https://doi.org/10.1007/s11033-019-05170-2

Asaduzzaman K, Mousumi T (2012) Nutritional and medicinal importance of *Pleurotus* mushrooms: an overview. Food Rev Intl 28:313–329. https://doi.org/10.1080/87559129.2011.637267

Belozerskaya TA, Gessler NN, Aver'yanov AA (2017) Melanin pigments of fungi. Fungal Metabol 263e291

Bhunia SK, Dey B, Maity KK, Patra S, Mandal S, Maiti S (2010) Structural characterization of an immunoenhancing heteroglycan isolated from an aqueous extract of an edible mushroom, *Lentinus squarrosulus* (Mont.) Singer. Carbohydr Res 345:2542–2549. https://doi.org/10.1016/j.carres.2010.09.015

Bisen PS, Baghel RK, Sanodiya BS, Thakur GS, Prasad GBKS (2010) *Lentinus edodes* : a macro-fungus with pharmacological activities. Curr Med Chem 17:2419–2430

Bohn JA, BeMiller JN (1995) (1→3)-β-D-glucans as biological response modifiers: a review of structure-functional activity relationships. Carbohydr Polym 28:3–14. https://doi.org/10.1016/0144-8617(95)00076-3

Butler MJ, Day AW (1998) Fungal melanins: a review. Can J Microbiol 44:1115–1136

Chang ST, Miles PG (1991) World production of edible mushrooms. Mushroom J 504:15–18

Charpentier BA, Sevenants MR, Sanders RA (1986) The shelf life of foods and beverages. Dev-Food-Sci Amsterdam: Elsevier Scientific Pub Co 12:413–433

Chen CC, Ho CT (1986a) High-performance liquid chromatographic determination of cyclic sulfur compounds of Shiitake mushroom (*Lentinus edodes* Sing.). J Chromatogr A 356:455–459

Chen CC, Ho CT (1986b) Identification of sulfurous compounds of shiitake mushroom (*Lentinus edodes* Sing.). J Agric Food Chem 34:830–833

Chen CC, Chen SD, Chen JJ, Wu CM (1984) Effects of pH value on the formation of volatiles of shiitake (Lentinus edodes), an edible mushroom. J Agric Food Chem 32:999

Chen W, Li W, Yang Y, Yu H, Zhou S, Feng J, Li X, Liu Y (2015) Analysis and evaluation of tasty components in the pileus and stipe of *Lentinula edodes* at different growth stages. J Agric Food Chem 63:795–801. https://doi.org/10.1021/jf505410a

Chen L, Gong Y, Cai Y, Liu W, Zhou Y, Xiao Y, Xu Z, Liu Y, Lei X, Wang G, Guo M, Ma X, Bian Y (2016) Genome sequence of the edible cultivated mushroom *Lentinula edodes* (Shiitake) reveals insights into lignocellulose degradation. PLoS One 11:e0160336. https://doi.org/10.1371/journal.pone.0160336

Chibata I, Okumura K, Takeyama S, Kotera K (1969) Lentinacin: a new hypocholesterolemic substance in *Lentinus edodes*. Experientia 25:1237–1238. https://doi.org/10.1007/bf01897467

Choi Y, Lee SM, Chun J, Lee HB, Lee J (2006) Influence of heat treatment on the antioxidant activities and polyphenolic compounds of Shiitake (*Lentinus edodes*) mushroom. Food Chem 99:381–387. https://doi.org/10.1016/j.foodchem.2005.08.004

Cronine DA, Ward MK (1971) The characterisation of some mushroom volatiles. J Sci Food Agric 22:477–479. https://doi.org/10.1002/jsfa.2740220912

Dermiki M, Phanphensophon N, Mottram DS, Methven L (2013) Contributions of non-volatile and volatile compounds to the umami taste and overall flavour of shiitake mushroom extracts and their application as flavour enhancers in cooked minced meat. Food Chem 141:77–83. https://doi.org/10.1016/j.foodchem.2013.03.018

Farook VA, Khan SS, Manimohan P (2013) A checklist of agarics (gilled mushrooms) of Kerala state, India. Mycosphere 4:97–131. https://doi.org/10.5943/mycosphere/4/1/6

Florence EJM, Yesodharan K (2000) Macrofungal flora of Peechi-Vazhani Wildlife Sanctuary. [KFRI Research Report no. 191.] Kerala: Kerala Forest Research Institute Peechi, Kerala, India, 43pp

Fujii T, Maeda H, Suzuki F, Ishida N (1978) Isolation and characterization of a new antitumor polysaccharide, KS-2 extracted from culture mycelia of *Lentinus edodes*. J Antibiot 31:1079–1090. https://doi.org/10.7164/antibiotics.31.1079

Gao S, Huang Z, Feng X, Bian Y, Huang W, Liu Y (2020) Bioconversion of rice straw agro-residues by *Lentinula edodes* and evaluation of non-volatile taste compounds in mushrooms. Sci Rep 10:1814. https://doi.org/10.1038/s41598-020-58778-x

Gavahian M, Cullen PJ (2019) Cold plasma as an emerging technique for mycotoxin-free food: efficacy, mechanisms, and trends. Food Rev Intl 36:193–214. https://doi.org/10.1080/87559129.2019.1630638

Gavahian M, Chu YH, Khaneghah AM, Barba FJ, Misra NN (2018) A critical analysis of the cold plasma induced lipid oxidation in foods. Trends Food Sci Technol 77:32–41. https://doi.org/10.1016/j.tifs.2018.04.009

Gavahian M, Khaneghah AM (2019) Cold plasma as a tool for the elimination of food contaminants: Recent advances and future trends. Crit Rev Food Sci Nutr 1–12. https://doi.org/10.1080/10408398.2019.1584600

Gavahian M, Chu YH, Jo C (2019a) Prospective applications of cold plasma for processing poultry products: benefits, effects on quality attributes, and limitations. Compr Rev Food Sci Food Saf 18:1292–1309. https://doi.org/10.1111/1541-4337.12460

Gavahian M, Sheu FH, Tsai MJ, Chu YH (2019b) The effects of dielectric barrier discharge plasma gas and plasma activated water on texture, color, and bacterial characteristics of shiitake mushroom. J Food Process Preserv 44:1–9. https://doi.org/10.1111/jfpp.14316

Gu K, Zhou CY, Shao Y (2016) Advances research and utilization on active constituents of edible fungus. Edible Fungus China 35:1–9

Guzman G, Salmones D, Tapia F (1997) *Lentinula boryana*: morphological variations, taxonomic position, distribution and relationships with *Lentinula edodes* and related species. Report Tottori Mycological Instit 35:1–28

Hibbett DS (1992) Towards a phylogenetic classification for shiitake [*Lentinus edodes*]: taxonomic history and molecular perspectives. Report Tottori Mycological Instit 30:30–42

Hibbett DS, Avenue D (2001) Shiitake mushrooms and molecular clocks: historical biogeography of *Lentinula*. J Biogeogr 28:231–241

Hibbett DS, Fukumasa-Nakai Y, Tsuneda A, Donoghue MJ (1995) Phylogenetic diversity in shiitake inferred from nuclear ribosomal DNA sequences. Mycologia 87:618–638. https://doi.org/10.2307/3760806

Hibbett DS, Hansen K, Donoghue MJ (1998) Phylogeny and biogeography of *Lentinula* inferred from an expanded rDNA dataset. Mycol Res 102:1041–1049. https://doi.org/10.1017/S0953756297005996

Hiraide M, Miyazaki Y, Shibata Y (2004) The smell and odorous components of dried shiitake mushroom, *Lentinula edodes* I: relationship between sensory evaluations and amounts of odorous components. J Wood Sci 50:358–364

Hoa HT, Wang CL (2015) The effects of temperature and nutritional conditions on mycelium growth of two oyster mushrooms (*Pleurotus ostreatus* and *Pleurotus cystidiosus*). Mycobiology 43:14–23. https://doi.org/10.5941/MYCO.2015.43.1.14

Ito Y, Toyoda M, Suzuki H, Iwaida M (1978) Gas-liquid chromatographic determination of lenthionine in shiitake mushroom (*Lentinus edodes*) with special reference to the relation between carbon disulfide and lenthionine. J Food Sci 43:1287–1289. https://doi.org/10.1111/j.1365-2621.1978.tb15289.x

Iwami K, Yasumoto K, Nakamua K, Mitsuda H (1975a) Agric Biol Chem 39:1933

Iwami K, Yasumoto K, Nakamua K, Mitsuda H (1975b) Agric Biol Chem 39:1941

Jiang Y (2011) Effects of extraction yield, structure, and properties of polysaccharides from *Lentinus edodes* by dynamic high-pressure microfluidization pretreatment. Nanchang University

Jiang T, Luo Z, Ying T (2014) Fumigation with essential oils improves sensory quality and enhanced antioxidant ability of shiitake mushroom (*Lentinus edodes*). Food Chem 172:692–698. https://doi.org/10.1016/j.foodchem.2014.09.130

Kała K, Kryczyk-Poprawa A, Rzewinska A, Muszynska B (2020) Fruiting bodies of selected edible mushrooms as a potential source of lovastatin. Eur Food Res Technol 246:713–722. https://doi.org/10.1007/s00217-020-03435-w

Kameoka H, Higuchi M (1976) The constituents of the steam volatile oil from *Lentinus edodes* Sing (*Corttinellus shiitake* P. Henn.) (in Japanese). Nippon Nougei Kagaku Gakkaishi 50:185–186

Kamiya T, Saito Y, Hashimoto M, Seki H (1969) Structure and synthesis of lentysine, a new hypo-cholesterolemic substance. Tetrahedron Lett 10:4729–4732

Keegan RH, Lu Z, Bogusz JM, Williams JE, Holick MF (2013) Photobiology of vitamin D in mushrooms and its bioavailability in humans. Dermato-Endocrinol 5:165–176

Kumar TKA, Manimohan P (2005) A new species of *Lentinus* from India. Mycotaxon 92:119–123

Leatham GF (1982) Cultivation of shiitake, the Japanese forest mushroom, on logs: a potential industry for the United States. For Prod J 32:29–35

Lee HY, Moon S, Shim D, Hong CP, Lee Y, Koo CD, Chung JW, Ryu H (2017) Development of 44 novel polymorphic SSR markers for determination of Shiitake mushroom (*Lentinula edodes*) cultivars. MDPI Genes 8:109. https://doi.org/10.3390/genes8040109

Liu JF, Peng JF, Chi YG, Jiang GB (2005) Determination of formaldehyde in shiitake mushroom by ionic liquid-based liquid-phase microextraction coupled with liquid chromatography. Talanta 65:705–709. https://doi.org/10.1016/j.talanta.2004.07.037

Liu Y, Yuan Y, Lei X, Yang H, Ibrahim SA, Huang W (2013) Purification and characterisation of two enzymes related to endogenous formaldehyde in *Lentinula edodes*. Food Chem 138:2174–2179. https://doi.org/10.1016/j.foodchem.2012.12.038

Maity S, Mandal EK, Maity K, Bhunia SK, Behera B, Maiti TK, Mallick P, Sikdar, Islam SS (2013) Structural study of an immunoenhancing polysaccharide isolated from an edible hybrid mushroom of *Pleurotus florida* and *Lentinula edodes*. Bioact Carbohydr Diet Fibre 1:72–80. https://doi.org/10.1016/j.bcdf.2013.01.003

Manimohan P, Divya N, Kumar TKA, Vrinda KB, Pradeep CK (2004) The genus *Lentinus* in Kerala state. Mycotaxon 90:311–318

Mason DJ, Sykes MD, Panton SW, Rippon EH (2004) Determination of naturally-occurring formaldehyde in raw and cooked Shiitake mushrooms by spectrophotometry and liquid chromatography–mass spectrometry. Food Addit Contam 21:1071–1082

Mattila P, Suonpaa K, Piironen V (2000) Functional properties of edible mushrooms. Nutrition 16:694–696. https://doi.org/10.1016/s0899-9007(00)00341-5

Mizuno T (1999) The extraction and development of antitumour-active polysaccharides from medicinal mushrooms in Japan. Int J Med Mushroom 1:9–29. https://doi.org/10.1615/IntJMedMushrooms.v1.i1.20

Mohanan C (2011) Macrofungi of Kerala. [KFRI Handbook no. 27.] Kerala: Kerala Forest Research Institute Peechi, Kerala, India, 670p. ISBN 81-85041-73-3

Moradali MF, Mostafavi H, Ghods S, Hedjaroude GA (2007) Immunomodulating and anticancer agents in the realm of macromycetes fungi (macrofungi). Int Immunopharmacol 7:701–724

Morita K, Kobayashi S (1967) Isolation, structure and synthesis of lenthionine and its analogs. Chem Pharm Bull 15:988–993. https://doi.org/10.1248/cpb.15.988

Muszynska B, Kala K, Wlodarczyk A, Krakowska A, Ostachowicz B, Gdula-Argasinska J, Suchocki P (2019) *Lentinula edodes* as a source of bioelements released into artificial digestive juices and potential anti-inflammatory material. Biol Trace Elem Res 194:603–613

Natarajan K, Kumaresan V, Narayanan K (2005) A checklist of Indian agarics and boletes (1984–2002). Kavaka 33:61–128

Nishikawa M, Kamiya K, Kobayashi S, Morita K, Tomiie Y (1967) The X-ray analysis of lenthionine, an odorous substance of shiitake, an edible mushroom. Chem Pharm Bull 15:756–760

Ohga S, Kitamoto Y (1997) Future of mushroom production and biotechnology. Food Rev Int 13:461–469

Okada S, Iga S, Isaka H (1972) Studies on formaldehyde observed in edible mushroom shiitake, *Lentinus edodes* (Berk.) Sing (in Giapponese). Jap J Toxicol & Environ Health 18:353–357

Pankaj SK, Bueno-Ferrer C, O'Neill L, Tiwari BK, Bourke P, Cullen PJ (2017) Characterization of dielectric barrier discharge atmospheric air plasma treated chitosan films: DBD plasma treatment of chitosan film. J Food Process Preserv 41(1):e12889. https://doi.org/10.1111/jfpp.12889

Pegler DN (1983) The genus *Lentinula* (Tricholomataceae tribe Collybieae). Sydowia 36:227–239

Pegler DN (1983a) The genus *Lentinus*: a world monograph. Kew Bulletin, Additional Series 10:1–281

Pegler DN (1983b) *Lentinus araucariae*, an Australasian member of the *Lentinus badius*-complex. Cryptogam Mycol 4:123–128

Poojary MM, Orlien V, Passamonti P, Olsen K (2017) Enzyme-assisted extraction enhancing the umami taste amino acids recovery from several cultivated mushrooms. Food Chem 234:236–244. https://doi.org/10.1016/j.foodchem.2017.04.157

Potluri S, Sangeetha K, Santhosh R, Nivas G, Mahendran R (2018) Effect of low-pressure plasma on bamboo rice and its flour. J Food Process Preserv 42(12):e13846. https://doi.org/10.1111/jfpp.13846

Pradeep CK, Vrinda KB (2007) Some noteworthy agarics from Western Ghats of Kerala. J Mycopathol Res 45:1–14

Przybylowicz P, Donoghue J (1988) The art and science of mushroom cultivation, shiitake growers handbook. Kendall/Hunt, Dubuque, pp 1–217

Raman V, Mohan NV, Balamuralitharan B, Rajendiran R, Rajangam V, Selvaraj A, Kim HJ (2019) Porous shiitake mushroom carbon composite with NiCo 2O4 nanorod electrochemical characteristics for efficient supercapacitor applications. Ionics 26(1):345–354. https://doi.org/10.1007/s11581-019-03178-z

Reis FS, Barros L, Martins A, Ferreira IC (2012) Chemical composition and nutritional value of the most widely appreciated cultivated mushrooms: an inter- species comparative study. Food Chem Toxicol 50:191–197

Rokujo T, Kikuchi H, Tensho A, Tsukitani Y, Takenawa T, Yoshida K, Kamiya T (1970) Lentysine: a new hypolipidemic agent from a mushroom. Life Sci 9:379–385

Saito M, Yamashita T, Kaneda T (1975) Quantitative analysis of eritadenine in "Shiitake" mushroom and other edible fungi. Eiyo to Shokuryo 28:503–505

Sakamoto Y, Nakade K, Sato S, Yoshida K, Miyazaki K, Natsume S, Konno N (2017) *Lentinula edodes* genome survey and postharvest transcriptome analysis. Appl Environ Microbiol 83(10):e02990–e02916

Samgina DI (1981) Agailkovie gribi 1. Agaricales. Fl Spor Rast Kazakhst 18:1–268

Senthilarasu G (2014) Diversity of agarics (gilled mushrooms) of Maharashtra, India. Curr Res Environ & Appl Mycol 4:58–78

Setiati Y, Chaidir L, Saputri A, Roosda AA (2019) Utilization of rice wastewater as an alternative medium for growth and quality of Shiitake mushrooms (*Lentinula edodes*). J Physic: Confer Series, 1402: 033034. IOP Publishing. https://doi.org/10.1088/1742-6596/1402/3/033034

Sharma JK, Mohanan C, Florence EJM (1985) Disease survey in nurseries and plantations of forest tree species grown in Kerala. KFRI Res Report 36:1–268

Shim D, Park SG, Kim K, Bae W, Lee GW, Ha BS, Ro HS, Kim M, Ryoo R, Rhee SK, Nou IS, Koo CD, Hong CP, Ryu H (2016) Whole genome de novo sequencing and genome annotation of the world popular cultivated edible mushroom, *Lentinula edodes*. J Biotechnol 223:24–25. https://doi.org/10.1016/j.jbiotec.2016.02.032

Shimomura N, Hasebe K, Fukumasa NY, Komatsu M (1992) Intercompatibility between geographically distant strains of Shiitake [*Lentinus edodes*]. Report Tottori Mycological Instit 30:26–29

Sun Y, Zhang M, Fang Z (2019) Efficient physical extraction of active constituents from edible fungi and their potential bioactivities : a review. Trends Food Sci Technol 105:468–482. https://doi.org/10.1016/j.tifs.2019.02.026

Suzuki H, Iiyama K, Yoshida O, Yamazaki S, Yamamoto N, Toda S (1990) Structural characterization of the immunoactive and antiviral water-solubilized lignin in an extract of the culture medium of *Lentinus edodes* mycelia (LEM). Agric Biol Chem 54(2):479–487

Tang LH, Jian HH, Song CY, Bao DP, Shang XD, Wu DQ, Tan Q, Zhang XH (2013) Transcriptome analysis of candidate genes and signaling pathways associated with light-induced brown film formation in *Lentinula edodes*. Appl Microbiol Biotechnol 97:4977–4989

Terashima K, Matsumoto T (2004) Strain typing of shiitake (*Lentinula edodes*) cultivars by AFLP analysis, focusing on a heat-dried fruiting body. Mycoscience 45:79–82

Tian Y, Zhao Y, Huang J, Zeng H, Zheng B (2016) Effects of different drying methods on the product quality and volatile compounds of whole shiitake mushrooms. Food Chem 197:714–722. https://doi.org/10.1016/j.foodchem.2015.11.029

Tokimoto K, Komatsu M (1978) In: Chang ST, Hayes WA (eds) The biology and cultivation of edible mushrooms, biological nature of *Lentinus edodes*. Academic Press, New York, pp 445–456

Varghese SP, Pradeep CK, Vrinda KB (2010) Mushrooms of tribal importance in Wayanad area of Kerala. J Mycopathol Res 48:311–320

Vetchinkina E, Kupryashina M, Gorshkov V, Ageeva M, Gogolev Y, Nikitina V (2017) Alteration in the ultrastructural morphology of mycelial hyphae and the dynamics of transcriptional activity of lytic enzyme genes during basidiomycete morphogenesis. J Microbiol 55:280–288

Wada S, Nakatani H, Morita K (1967a) A new aroma-bearing substance from shiitake, an edible mushroom. J Food Sci 32:559–561

Wada S, Nakatani H, Fujinawa S, Kimura H, Hagaya M (1967b) Studies on lenthionine, a new aroma-bearing substance from shiitake (part 1) (in Japanese). J Jap Soc Nutrit & Food Sci 20:355–359

Wasser SP (2002) Medicinal mushrooms as a source of antitumor and immunomodulating polysaccharides. Appl Microbiol Biotechnol 60:258–274

Wasser SP (2003) In: Coates PM (ed) Shiitake (Lentinus edodes). Encyclopedia of dietary supplements. Marcel Dekker, New York, pp 653–664

Wasser SP, Weis AL (1997) In: Nevo E (ed) Medicinal Mushrooms, Lentinus edodes (Berk.) Singer. Peledfus Publ. House, Haifa, p 95

Weng X, Chon CH, Jiang H, Li D (2009) Rapid detection of formaldehyde concentration in food on a polydimethylsiloxane (PDMS) microfluidic chip. Food Chem 114:1079–1082

Whiteford JR, Thurston CF (2000) The molecular genetics of cultivated mushrooms. Adv Microb Physiol 42:1–23

Wu X, Li H, Zhao W, Fu L, Peng H, He L, Cheng J, Wei H, Wu Q (2010) SCAR makers and multiplex PCR-based rapid molecular typing of Lentinula edodes strains. Curr Microbiol 61:381–389

Xu X, Yan H, Tang J, Chen J, Zhang X (2014) Polysaccharides in Lentinus edodes: isolation, structure, immunomodulating activity and future prospective. Crit Rev Food Sci Nutr 54:474–487. https://doi.org/10.1080/10408398.2011.587616

Xu Y, Tian Y, Ma R, Liu Q, Zhang J (2016) Effect of plasma activated water on the postharvest quality of button mushrooms, Agaricus bisporus. Food Chem 197:436–444. https://doi.org/10.1016/j.foodchem.2015.10.144

Xu Z, Meenu M, Xu B (2019) Effects of UV-C treatment and ultrafine-grinding on the biotransformation of ergosterol to vitamin D2, physiochemical properties, and antioxidant properties of shiitake and Jew's ear. Food Chem 309:125738. https://doi.org/10.1016/j.foodchem.2019.125738

Yamanaka K (1997) Production of cultivated edible mushrooms. Food Rev Intl 13:327–333

Yamazaki H, Ogasawara Y, Sakai C, Yoshiki M, Makino K, Kishi T, Kakiuchi Y (1980) Formaldehyde in Lentinus edodes (in Giapponese). J Food Hygiene Soc Japan 21:165–170

Yan D, Liu Y, Rong C, Song S, Zhao S, Qin L, Wang S, Gao Q (2020) Characterization of brown film formed by Lentinula edodes. Fungal Biol 124(2):135–143. https://doi.org/10.1016/j.funbio.2019.12.008

Yasumoto K, Iwami K, Mitsuda H (1971) A new sulfur-containing peptide from Lentinus edodes acting as a precursor for lenthionine. Agric Biol Chem 35:2059–2069. https://doi.org/10.1080/00021369.1971.10860187

Yasumoto K, Iwami K, Mitsuda H (1976) Enzymic formation of Shiitake aroma from non-volatile precursor(s)-lenthionine from lentinic acid. Mushroom Sci 9:371–383

Yoo S-I, Lee H-Y, Markkandan K, Moon S, Ahn YJ, Ji S, Ko J, Kim S-J, Ryu H, Hong CP (2019) Comparative transcriptome analysis identified candidate genes involved in mycelium browning in Lentinula edodes. BMC Genomics 20:121. https://doi.org/10.1186/s12864-019-5509-4

Yu Z, Ming G, Kaiping W, Zhixiang C, Liquan D, Jingyu L, Fang Z (2010) Structure, chain conformation and antitumor activity of a novel polysaccharide from Lentinus edodes. Fitoterapia 81:1163–1170. https://doi.org/10.1016/j.fitote.2010.07.019

Zhang M, Cui SW, Cheung PCK, Wang Q (2007) Antitumor polysaccharides from mushrooms: a review on their isolation process, structural characteristics and antitumor activity. Trends Food Sci Technol 18:4–19. https://doi.org/10.1016/j.tifs.2006.07.013

Zhang Y, Xia L, Pang W, Wang T, Chen P, Zhu B, Zhang J (2013) A novel soluble β-1,3-D-glucan salecan reduces adiposity and improves glucose tolerance in high-fat diet-fed mice. Br J Nutr 109(2):254–262

Zhuang H, Chen Z, Feng T, Yang Y, Zhang J, Liu G, Li Z, Ye R (2017) Characterization of Lentinus edodes β-glucan influencing the in vitro starch digestibility of wheat starch gel. Food Chem 224:294–301. https://doi.org/10.1016/j.foodchem.2016.12.087

Chapter 11
Fungi: A Potential Future Meat Substitute

Meganathan Bhuvaneswari and Nallusamy Sivakumar

11.1 Introduction

In the past century, worldwide agriculture production was significantly improved to provide food supply to meet the demands of the rapidly growing population. Just 10% increase of the cultivable land area results in a doubled production (FAO 2010). But the most drastic parts are the population explosion, urbanization, lifestyle change, poverty, health conditions, and dietary requirements which affect the global agricultural production (Augustin et al. 2016). Food industry, one of the largest exploiters of environment and resources, leads to severe loss of biodiversity and degradation of ecosystem (Dunne et al. 2002; Tscharntke et al. 2012; Röös et al. 2013). Most of the water footprint, around 70–85%, is mainly from human activities toward agriculture. Even around 30% of the total greenhouse gases were emitted as a side product of the agricultural sector (Smetana et al. 2015). Livestock production accounts for the largest emitter of methane. Having meat in our diet is nowadays unfriendly for the environment because of the inefficient usage of land, water, and energy and also for the amount of unwanted gases released through meat production (Pimentel and Pimentel 2003, McMichael et al. 2007).

Finding meat substitutes or replacers and commercializing them is still in infancy stage since the market share is just 1–2% of the total meat market. The consumers of meat substitutes are influenced by the nutritional effect of the substitute and its environmental impact. In the middle of this century, the global population is about to decelerate its growth and may reach the plateau of nine billion which will exert an extra pressure to feed this population (Godfray et al. 2010). Thus, the main future

M. Bhuvaneswari
Department of Biotechnology, Sona College of Arts and Science, Salem, India

N. Sivakumar (✉)
Department of Biology, College of Science, Sultan Qaboos University, Muscat, Sultanate of Oman

X. Dai et al. (eds.), *Fungi in Sustainable Food Production*, Fungal Biology, https://doi.org/10.1007/978-3-030-64406-2_11

181

challenges have to be addressed immediately to improve the public health by improving and providing healthy food and nutritional security. In this scenario, the following questions are arising: whether a suitable meat substituent can be identified? If at all identified whether it can be a complete alternative for the meat? Will people recognize and accept these alternatives? This book chapter will answer few of these questions, and the prime answer will be fungal mycoproteins.

11.2   Importance of Meat

Worldwide, nonvegetarian consumers prefer meat as their first choice of consumption. This is because the meat is a tasty, juicy, flavor-rich, chewy-textured food with high nutritional value. Justification of meat consumption mainly relies on Ns, i.e., it is natural, normal, necessary, and nice (Piazza et al. 2015). Moreover, people rely on meat for its taste and variety (Asgar et al. 2010). It is considered as the food with the highest protein source and high biological values of vitamins and minerals. The role of proteins present in the meat is having two main features: (i) The meat protein contains all essential amino acids more or less closely resembling the needs of human body, making it as more nutritious and preferable, and (ii) the functional properties of meat proteins cannot be reproduced by any other food proteins in the food industry (Xiong 2004).

Food industry also produces various processed meats such as slathered and marinated meat, restructured meat rolls and loaves, luncheon meats, boneless ham, sausage, frankfurters, 30% fat bologna, etc. The food industries use not only the muscle meat but also the other parts of the animals to produce products such as sausages, hams, and bologna which have high level of fat (Roth et al. 1997).

11.3   Health Concerns of Meat

Although meat is a preferred food, it is also having some negative aspects such as high calorie and high fat content which impacts the human health. The main demerit of the consumption of red meat is its increased rate of mortality. This has been reported by several studies (Pan et al. 2012; Rohrmann et al. 2013). Moreover, it is environmentally unfriendly by its inefficient land use and also the level of greenhouse gases it emits (McMichael et al. 2007). Green activists, People for the Ethical Treatment of Animals (PETA) like organizations, and some religious ritualistic features also condemn the use of animals for the meat production. Policy makers are also in association with these organizations which emphasize customers to make a shift toward sustainable products (Malav et al. 2015).

Even though the nutritional value of meat makes it as a prominent food source, excessive intake of meat is not recommended due to its high fat content (Muguerza et al. 2004, Cengiz and Gokoglu 2005). Apart from essential amino acids, vitamins, and minerals, meat consists of higher proportion of saturated fatty acids than poly-unsaturated fatty acids (PUFAs). The PUFAs such as n-3 PUFAs and α-linolenic acid (ALNA C18:3) are capable of reducing pathogenesis of many diseases such as cancer, cardiovascular disease (CVD), and autoimmune and inflammatory diseases. These n-3 PUFAs and α-linolenic acid are present in plant proteins than animal proteins (Jimenez-Colmenero 2007). The WHO prescribes a diet for CVDs patients in which fat must be between 15% and 30% of the calories, saturated fat must be less than 10%, and intake of cholesterol must be less than 300 mg/day (WHO 2003). But the consumption of meat on a regular basis by normal persons and elderly persons and for CVDs patients impacts their health badly.

11.4 Sustainable Diet

The concept of sustainable diet is to eliminate poverty and also reduce its impact in the environment. The diet must also eradicate poor health outcomes and the nutritional insecurity (Johnson et al. 2014). The FDA also suggests the same by naming it as climate-smart agriculture through which it combats food security and environment (FAO 2010). The above-said guidelines can be followed by reducing meat consumption and livestock production which utilizes the land resources, the largest emitter of methane as it leads to land degradation and deforestation. Thus researchers are looking for a replacement of meat with some other food products, with less harm to the environment. The products which come under these are meat substitutes, meat analogs, meat replacers, and meat alternatives (Hoek et al. 2011).

The main sources of the alternatives can be plants (soy, pea, oat), animals (milk, insects), or microbial products (mycoproteins) (Smetana et al. 2015). The ethical and environmental concerns made this meat substitute industries to expand rapidly, and the market also predicted to have a turnover of $6 billion in 2022 (Ritchie et al. 2017; Godfray et al. 2018). The success of the meat substitute industry is because of the nutritional effects it brings in the diet. Replacing meat in a diet is always easy when the meat substitutes fit the taste and nutritional effect. The most important is to establish the popularization of meat substitutes and also to enhance the public awareness about the dietary impacts.

11.5 Purpose of Searching for Meat Alternatives

Replacement for meat by substitutes has existed for many years for halal and kosher markets. The meat protein replacement has gained interest particularly from the end of the last century due to emergence of many diseases to the livestock such as mad cow disease and foot-and-mouth disease. This leads to much concern about using meat protein from those infected animals. Religious concerns also lead to the prohibition of meat and meat products. The major religions such as Islam, Jews, and Hindus have their own choices for their meat consumption.

High prices of meat and meat products also account to search for meat alternatives. Lower prices of vegetables proteins than muscle proteins will definitely reduce the cost of the meat products. Moreover, the ready acceptance of vegetable proteins such as textured soy protein (TSP) is due to its low cost (Singh et al. 2008). Nearly 800 million malnourished people are there in the underdeveloped countries (Myers 2002). Availability of nutritious food for the poor and malnourished is always a great challenge for the developing countries (Bhat and Karim 2009, Boye et al. 2010). Thus various factors such as animal diseases, global shortage of animal proteins, demand for cholesterol-free, low saturated fat food searches (Franklin 1999; Maurer 2002), as well as the religious concerns, economic reasons, environmental impact, pressure to avoid meat, rising vegetarianism, and animal rights movement necessitate the look for a meat substitute.

11.6   Artificial Meat

The meat industry struggles to cope up with the increasing demand for the meat. Hence, artificial meat gains much attraction in the recent years. This leads to the development of nontraditional meat and protein products. These novel meat and protein products are called artificial meat. The development of artificial meat such as in vitro or cultured meat and meat from genetically modified organisms is still at the early stage. However, these would become tough competitors to the conventional meat production in the future.

Bonny et al. (2015) categorized the artificial meat into three different types. The first one is meat alternative which refers to alternative protein sources from plants and fungi. The second type is the tissue and cells cultured in the labs called cultured meat or in vitro meat, and the third category of artificial meat is from genetically modified organisms.

Meat extenders can be flakes, in minced form (>2 mm) and in chunk form (15–20 mm) which can absorb 3–5 times of water to their original weight (Riaz 2004). Food researchers have formulated "meat analogs" which can satisfy the consumers. Analogs can be defined as structurally similar but differs in composition. The meat analogs are products which resemble meat in appearance, color, flavor, and texture. Meat analogs can be of any size and made into sheets, disks, patties, strips, and other shapes too. These analogs have a striated, layered structure and may be called as meat substitutes, mock meat, faux meat, or imitation meat (Sadler 2004). Though they possess the same qualities of meat, they are inexpensive.

11.7 Types of Nonmeat Proteins and Their Limitations

Due to many limitations and rationales of animal protein, the meat alternative sources or meat supplements are in demand. Normally plant proteins and fungal proteins are in the race for developing meat substitutes. Plant proteins which are promising as meat substitutes are soy proteins, legume proteins, oilseed proteins, and cereal proteins. Subsequently fungal proteins (mycoproteins) are gaining momentum for developing meat substitutes, along with the above-said plant proteins.

11.7.1 Soy Proteins

Glycine max – soybean – is a traditional food crop belonging to leguminous plants. The leguminous plants are known for their protein content, and this soybean in particular is a rich dietary source for proteins. Various food products are available from soy like soy milk (Fukushima 1994), tofu, and tempeh, a fermented soy product of Indonesian origin. Soybeans' nutritional value is calculated as 35–40% of proteins (high-quality proteins with balanced amino acids), 15–20% of fat, 30% of carbohydrates, and 10–30% moisture. It is also rich in fiber, iron, calcium, zinc, and vitamins, which makes it as a very special meat alternative. The main disadvantage of soya proteins is their strong flavor of the products. The flavors such as grass and bean flavor and the bitter and astringent flavor reduce the taste of the meat alternatives. Their off-flavors intensify in the stepwise processing (Okubo et al. 1992). Heat treatment and germination can remove this objectionable odor and flavor (Suberbie et al. 1981). The allergic reactions caused by consumption of soy proteins are also a main disadvantage for developing meat substitutes.

11.7.2 Legume Proteins

Legume proteins are the edible seeds of leguminous plants which includes beans and pulses (Oboh et al. 2009). These seeds supply a high amount of protein. These legume seeds are easily available even for the economically poor people, and so it is also called as poor man's meat. The amount of protein accounts to 20–30% in its total dry weight. The underdeveloped and developing countries always try to improve the nutritional content and the functional properties.

Main disadvantages of the legume proteins are the low amount of sulfur-containing amino acids and the allergenic food products in many legume crops such as peanuts, soybeans, lentils, beans, chickpeas, tree nuts, etc. (Sing et al. 2008, Riascos et al. 2010). Moreover, the legume seeds are inherited with protein inhibitors to resist and to protect itself from degradation, and this resistant power inhibits the digestive enzymes and so affects the digestive process in humans when consumed (Leterme et al. 1992).

11.7.3 Oilseed Proteins

The oil-producing plants will be having high level of proteins which is useful for the human diet. The oilseed proteins can be categorized into biologically active proteins (enzymes), storage proteins, and structural proteins. Among these, the storage proteins are the one which is very much abundant in the oilseeds such as 25% for rapeseed, 23% for cottonseed, and 13–17% for safflower (Asgar et al. 2010). The main disadvantages of the oilseed protein alternative are antinutrients which are very much necessary for the safe protection of oilseeds but reduce the enzymes in human digestion. The allergenic properties of the oilseeds are also to be considered before deciding the meat alternatives.

11.7.4 Cereal Proteins

Worldwide cereals are the most important food crops, and so their products are also most important. They are consumed as (i) seeds, rice, barley, oats, and maize; (ii) flour, rice, wheat, and rye; and (iii) flakes, barley, oats, and maize. Wheat has the highest protein content of all (8–17.5%) followed by maize (8.8–11.9%), barley (7–14.6%), rice (7–10%), oats (8.7–16%), and rye (7–14%). The main proteins are albumins, globulins, gliadins, and glutelins (Guerrieri 2004). The most promising member of cereal protein for meat analog is the wheat gluten. Combined with soy flour, wheat gluten produces meat extenders for meat patties and binder for sausage products (Riaz 2004). The main disadvantage is the gluten, the protein found in wheat, rye, and barley which are harmful to some individuals. Intolerance of gluten proteins can lead to celiac disease which is not a normal hypersensitive reaction mediated by IgE but affects and damages the small intestine (Sadler 2004). The water insolubility is also one major impact which limits its use as meat analog (Kong et al. 2007).

11.8 Mycoproteins

Consumers need a meat alternative which does not need to resemble meat in texture, taste, and flavor, but it needs to be like as such as meat itself. In this scenario, the substitution of mycoproteins as meat substitute is accepted easily than the plant-rich proteins. This is because the mycoproteins are more similar to meat and so the consumers are eased with this substitution (Raats 2007).

11.9   Mycoprotein as a Meat Substitute

Mycoproteins are the proteins obtained from filamentous fungi that can be utilized for human consumption. Already the members of mushrooms and truffles are consumed as food because of its pleasant taste and health benefits (Boland et al. 2013). But they are not considered as meat alternatives because of their lower protein content. Thus in search of the high protein content fungal members, the filamentous fungal members are preferred because of their high growth and rich protein nature. Moreover, these fungi have been consumed by human beings for a long time to improve the nutritional value and as a component of fermented food (Nout and Aidoo 2002). Different fungal members are used to produce different food and fermentative products. On the other hand, the fungal biomass formed by filamentous fungi growth, a rich source of mycoprotein, can be processed and used as human food. In 2002, the FDA declared that mycoproteins are generally recognized as safe (GRAS) (Denny et al. 2008).

The fibrous texture of the mycelium, which is comparable to the texture of the end product the meat, makes the filamentous fungi as the suitable candidate for the production of meat substitute (Edelman et al. 1983). Often the mycoproteins from fungal mycelium are added with other agents such as albumin from egg to obtain the texture which is almost similar to that of meat (Denny et al. 2008, Rodger 2001). Meat muscle cells are held together by a connective tissue, and the mycoproteins can be bound with albumin which can bind hyphae and can form suitable texture like meat. The fibrous texture of the mycelia of the filamentous fungus is comparable to that of meat and to its final products which makes it into a meat analog.

11.10   *Fusarium venenatum*

Globally there was a pressure on food scientists to develop microbial proteins to combat protein shortage. In the year 1967, a most versatile member of filamentous fungi *Fusarium venenatum* was identified and used for mycoprotein production (Denny et al. 2008). The name mycoprotein is a generic name given for the fungal biomass mainly having filamentous hyphae, when the RNA content was reduced during the fermentation process. In search of a viable fungal isolate, around 3000 fungal isolates were analyzed globally, and then this *F. venenatum* A3/5 (ATCC PTA-2684) is found as the best organism for mycoprotein production (Wiebe 2002).

11.11   Production of Mycoproteins

Continuous flow fermentation of *F. venenatum* using glucose substrate is the most preferred method of mycoprotein production (Denny et al. 2008). Many parameters have to be considered such as CO_2 evolution rate, optimal temperature (28–30 °C), optimum pH (6.0), and time duration (6 weeks).

The concentration of biomass determines the CO_2 evolution state and so the flow rate of the medium (Wiebe 2002). The production of mycotoxins can be checked every 6 h to ensure that the produced mycoproteins are toxin-free. While harvesting the fungal biomass, it has to be kept for short-time heat treatment to get rid of RNA, by activating cells' endogenous RNAse enzymes. The temperature chosen for the RNA treatment is 68 °C for 30–45 min. Once this process is over, the mycoproteins can be isolated from the fungal broth by centrifugation process as a paste. This treatment will reduce the RNA content to 2% from 10% of its dry weight (Asgar et al. 2010).

11.12   Nutritional and Health Benefits

The mycoproteins are a low-fat, high-protein, and high-fiber food component. It also contains a minimal amount of sodium and enriched amount of zinc, selenium, and antioxidants (Denny et al. 2008; Smith et al. 2015). The chemical composition of fungal mycelium provides very good nutritional benefits. The mycelial cell wall, membrane, and cytoplasm serve as sources of dietary fiber, polyunsaturated fatty acids, and high-quality proteins, respectively (Fig. 11.1). Each part of the fungal cell is an incredible healthy component (Rodger 2001). It is highly comparable with vegetable protein sources by its cholesterol-free content, low unsaturated fatty acid profile, high fiber content, etc. The fiber present in the mycoprotein such as chitin and β-glucan forms a matrix of chitin-glucan which is not water soluble (88% insolubility) (Bottin et al. 2016). The chitin present here amounts to one-third in proportion, and it is not commonly present in human diet, whereas the remaining two-thirds is the β-glucan subunits.

The study of Sadler (2004) states that the chitin ingestion helps to get relief of joint pain due to osteoarthritis and also stimulates the growth of colonization of beneficial bacteria in the colon. Various studies of these mycoproteins reveal that these fibers have improved the glycemic profile and also increase the insulin

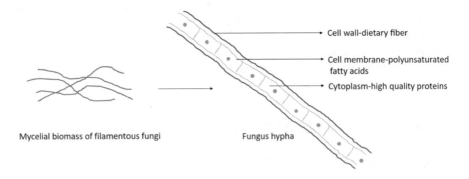

Fig. 11.1 The fugal mycelium and its nutritional benefits

sensitivity (Bottin et al. 2016, Denny et al. 2008, Turnbull et al. 1992). The more recent study compared the health aspects of mycoprotein and milk which found that mycoprotein is having longer hyperinsulinemia. Insulin shows a 45-min peak for mycoprotein meal whereas 15-min peak for milk (Dunlop et al. 2017). The most important aspect of consumption of mycoprotein is the lowering of blood cholesterol level when tested for those with raised cholesterol levels (Ruxton and McMillan 2010).

For the last two decades, our diet is changing rapidly, but the standard dietary plan with major and minor nutrient levels is very much necessary. Among those nutrients the protein intake in a required quantity is more desirable, i.e., 15% of the total energy intake (1.1 g of protein/kg of body weight). Elderly people require 1.2 g of proteins/kg of wt (Nordic Council of Ministers et al. 2014, Filho et al. 2019). If adequate quantity is not taken, it may lead to edema, detrimental changes in hair and skin, as well as muscle weakness. Diseases like kwashiorkor and marasmus may occur due to protein-energy malnutrition (Batool et al. 2015). Thus the mycoproteins, a rich source of proteins, are necessary for all sorts of persons. When compared to other proteins, the mycoproteins show satiety due to protein content and fiber content in the mycoproteins and presence of all essential amino acids (Williamson et al. 2006, Paddon et al. 2008, Slavin and Green 2007, Bottin et al. 2016). Mycoprotein diet when taken will facilitate weight loss, reduce energy intake, and fight hunger (Filho et al. 2019). The use of mycoproteins as a main ingredient for many products was taken as a research for nearly two decades. The results were promising, and it was readily accepted by the European and American food markets. Including mycoproteins in diets gives a tasty and healthy satisfaction for the consumers (Fig. 11.2). Moreover, if it is included in the diet on a regular basis, the people can easily manage obesity, diabetes, arthritis, heart diseases, etc. (Asgar et al. 2010).

11.13 Environmental Aspects

Artificial meat products may have an advantage over conventional meat production by emitting less amount of greenhouse gases. It is estimated that 18% of all greenhouse gas emissions were produced by livestock (FAO 2009). However, many works reported lower level of greenhouse gas emissions by livestock (Pitesky et al. 2009). On the other hand, any healthy substitute for meat would reduce the farming ruminant animals grown for the meat production which in turn reduces the greenhouse gas emissions. Hence, the mycoprotein as a meat substitute will be helpful for maintaining the environmental susceptibility. To get a clear idea, more studies on full life cycle analysis should be conducted. It was calculated that in vitro meat production may reduce 99% of land use, 82–96% of water usage, 78–96% of greenhouse gas emissions, and 7–45% of energy consumption (Tuomisto et al. 2011). Hence, the environmental impact of cultured meat is significantly lower than the conventional meat. However, production of cultured meat on an industrial scale

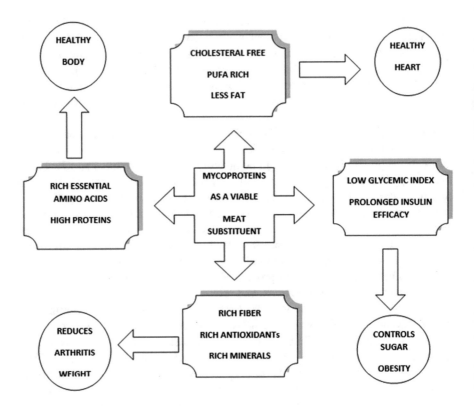

Fig. 11.2 Nutritional and health benefits of mycoproteins

needs more technology, cost, and time and remains challenging. The environmental impact of mycoprotein produced using sugar-based molasses was compared with other meat options. It was found that mycoprotein and chicken have almost indistinguishable environmental impact which is lower than the cultured meat (Smetana et al. 2015).

They compared the various environment impact factors such as energy, land water used for the meat substitutes, and the real meat members. This study revealed that mycoproteins stood as an efficient meat substitute in managing the above-said factors (Table 11.1). Mycoprotein is considered as an efficient alternative in terms of water (~500 L/kg) and land use. When considering the land usage, for chicken production 5–7 m^2a/kg and pork production 7–8 m^2a/kg of land are needed. However, mycoprotein needs only <2 m^2a/kg which is very lower than chicken and pork. In energy consumption mycoproteins are on par with dairy alternatives (15–20 kWh/kg). However, it is higher than vegetables and insects (Smetana et al. 2018; Filho et al. 2019). In addition, it was reported that when compared to cultured meat, the mycoprotein performance is lower with reference to calorific energy value and the digestible proteins (Smetana et al. 2015).

Table 11.1 Environmental impact of different meat products

S. No.	Environmental factors	Mycoprotein	Chicken	Pork	Sustainable mycoprotein production
1	Land	<2 m²a/kg	5–7 m²a/kg	7–8 m²a/kg	0.5 m²a
2	Water	~500 L/kg	More usage	More usage	~250 L/kg
3	Energy	~15–20 kwh/kg	More usage	More usage	~10 kwh/kg

Mycoproteins can be produced by utilizing the agro-industrial residues which can decrease the environmental impact. The synthesis and production can directly use lignocellulosic materials without pretreatment in submerged culture as well as in solid-state fermentation (Satari and Karimi 2018). When the agro-industrial residues are used, then the production will still has less environmental impact, for per kg production of mycoprotein. But the large-scale trials still need clear safety measures, production strategies, and regulatory measures.

11.14   Future of Meat Substitutes

Though the food manufacturers are trying to bring soy-, wheat-, and pea-based products as meat substitutes in the market, the future will depend on the new meat alternative from fungi. According to the food business news, there are two companies going to launch their first meat alternative in the restaurants of Colorado in 2020. Fungi have several advantages over conventional plant-based proteins. The fibrous texture of fungal products permits a wide variety of types and cuts of meal. The most important characteristic feature of fungal meat is its neutral test which abolishes the use of masking agents. As discussed already, the fungal meat production needs a little bit of land and less amount of water and energy with packed nutrients which makes it a better alternative and competitor for plant-based meat products. Meat products of fungal origin are produced in the laboratories using the vegetative threadlike structures called mycelium. The fungi were grown by the companies in large tanks by supplying the nutrients. After growth the mycelia were harvested, cut, and flavored to match the taste of meat. Fine-tuning of the final product can be made by adjusting the fat ratios. Fungal meats are the best alternative to plant-based products because they lose some nutrients present in the plants during the processing (https://www.foodbusinessnews.net/articles/15226). The biggest challenge is educating the consumers on the benefit of consuming fungi-based meats over plant-based and the conventional meat.

11.15   Limitations

Mycoproteins may be mildly allergic to some persons (Denny et al. 2008). Mycoproteins have a lower amount of vitamin B12 and iron content than meat. But the major concern about the mycoprotein is its high RNA content. Nearly 10% of fungal biomass (dry weight) is of RNA, which is comparatively much higher than muscle meat (less than 0.6%) and edible oats (0.6–2%) (Jonas et al. 2001). This higher amount of RNA always leads to increased amount of uric acid which is a bigger problem for gout (Denny et al. 2008; Jonas et al. 2001). This high level of RNA can be overcome by rapid heat treatment of the fungal biomass at 68 °C for 20–45 min which reduces the RNA content to less than 2%. The RNAs are broken down to monomers, and they will be leaving out of the cells by diffusion (Raats et al. 2007).

11.16   Conclusion

Globally the food market has become more complex, and the search of meat alternatives is at great pace. The challenge of providing balanced nutritious, healthy, and affordable foods with similar taste and texture has been answered via mycoproteins as meat substitutes. Consumers choose meat alternative products for their nutritional benefits and also for their resemblance of meat in texture and taste. Mycoprotein which is considered as a powerhouse component with less fat with less calories, more fiber, and good taste and texture can be a promising substitute of the future. Though several health and nutritious benefits are available, unfortunately these mycoproteins are quite unpopular. Many countries are unaware of these mycoproteins and their health benefits. Mycoproteins have proven positive impact as a meat substitute, but the challenge is producing these proteins in a cost-effective manner with a sustainable production methodology. Solid-state fermentation and any other production process using industrial or agro-residues can be an alternative production process. Thus it is a tough task to popularize these mycoproteins as an alternative protein source and as a healthy meat substitute. A mighty approach to popularize and to make these mycoproteins and their products familiar, campaigns are necessary to enhance the consumer knowledge. Technological developments toward the production and fetching consumer awareness and popularity will make mycoproteins as a viable substitute in the near future, and these measures can establish them as effective meat substitutes.

References

Asgar MA, Fazilah A, Huda N, Bhat R, Karim AA (2010) Nonmeat protein alternatives as meat extenders and meat analogs. Compr Rev Food Sci Food Safety 9:513–529

Augustin MA, Riley M, Stockmann R, Bennett L, Kahl A, Lockett T, Osmond M, Sanguansri P, Stonehouse W, Zajac I, Cobiac L (2016) Role of food processing in food and nutrition security. Trends Food Sci Technol 56:115–125

Batool R, Butt MS, Sultan MT, Saeed F, Naz R (2015) Protein-energy malnutrition: a risk factor for various ailments. Crit Rev Food Sci Nutr 55:242–253

Bhat R, Karim AA (2009) Exploring the nutritional potential of wild and underutilized legumes. CRFSFS 8:305–331

Boland MJ, Rae AN, Vereijken JM, Meuwissen MPM, Fischer ARH, van Boekel MAJS, Rutherford SM, Gruppen H, Moughan PJ, Hendriks WH (2013) The future supply of animal-derived protein for human consumption. Trends Food Sci Technol 29:62–73

Bonny SPF, Gardner GE, Pethick DW, Hocquette JF (2015) What is artificial meat and what does it mean for the future of the meat industry? J Integr Agric 14(2):255–263

Bottin JH, Swann JR, Cropp E, Chambers ES, Ford HE, Ghatei MA, Frost GS (2016) Mycoprotein reduces energy intake and post-prandial insulin release without altering glucagon-like peptide-1 and peptide tyrosine-tyrosine concentrations in healthy over-weight and obese adults: a randomised-controlled trial. Br J Nutr 116:360–374

Boye J, Zare F, Pletch A (2010) Pulse proteins: processing, characterization, functional properties and applications in food and feed. Food Res Int 43:414–431

Cengiz E, Gokoglu N (2005) Changes in energy and cholesterol contents of frankfurter-type sausages with fat reduction and fat replacer addition. Food Chem 91:443–447

Denny A, Aisbitt B, Lunn J (2008) Mycoprotein and health. Nutr Bull 33:298–310

Dunlop MV, Kilroe SP, Bowtell JL, Finnigan TJA, Salmon DL, Wall BT (2017) Mycoprotein represents a bioavailable and insulino-tropic non-animal-derived dietary protein source: a dose–response study. Br J Nutr 118:673–685

Dunne JA, Williams RJ, Martinez ND (2002) Network structure and biodiversity loss in food webs: robustness increases with connectance. Ecol Lett 5:558–567

Edelman J, Fewell A, Solomons (1983) Myco-protein—a new food. Nutr Abstr Rev 53:472–479

FAO (2009) The state of food and agriculture. Electronic publishing policy and support branch. Communication Division, FAO, Rome

FAO (2010) 'Climate-smart' agriculture: policies, practices and financing for food security, adaptation and mitigation. Food and Agriculture Organization, Rome

Filho PFS, Andersson D, Ferreira JA, Taherzadeh MJ (2019) Mycoprotein: environmental impact and health aspects. World J Microbiol Biotechnol 35:147

Franklin A (1999) Animals and modern cultures: a sociology of human—animal relations in modernity. Sage, London

Fukushima D (1994) Recent progress on biotechnology of soybean proteins and soybean protein food products. Food Biotechnol 8:83–135

Godfray HCJ, Beddington JR, Crute IR, Haddad L, Lawrence D, Muir JF, Pretty J, Robinson S, Thomas SM, Toulmin C (2010) Food security: the challenge of feeding. Billion People Sci 327:812–818

Godfray HCJ, Aveyard P, Garnett T, Hall JW, Key TJ, Lorimer J, Pierrehumbert RT, Scarborough P, Springmann M, Jebb SA (2018) Meat consumption, health, and the environment. Science 361:1–10

Guerrieri N (2004) Cereal proteins. In: Yada RY (ed) Proteins in food processing. Woodhead Publishing Limited, London, pp 176–196

Hoek AC, Luning PA, Weijzen P, Engels W, Kok FJ, de Graaf C (2011) Replacement of meat by meat substitutes. A survey on person- and product-related factors in consumer acceptance. Appetite 56:662–673

https://www.foodbusinessnews.net/articles/15226-is-fungi-based-protein-the-future-of-fake-meat. Accessed on 10 May 2020

Jiménez-Colmenero F (2007) Healthier lipid formulation approaches in meat-based functional foods. Technological options for replacement of meat fats by nonmeat fats. Trends Food Sci Technol 18:567–578

Johnston JL, Fanzo JC, Cogill B (2014) Understanding sustainable diets: a descriptive analysis of the determinants and processes that influence diets and their impact on health. Food Secur Environ Sustain Adv Nutr 5:418–429

Jonas DA, Elmadfa I, Engel KH, Heller KJ, Kozianowski G, König A, Müller D, Narbonne JF, Wackernagel W, Kleiner J (2001) Safety considerations of DNA in food. Ann Nutr Metab 45:235–254

Kong X, Zhou H, Qian H (2007) Enzymatic preparation and functional properties of wheat gluten hydrolysates. Food Chem 101:615–620

Leterme P, Monmart T, Thrwes A (1992) Varietal distribution of trypsin inhibitors in peas. In: Picard J (ed) Proceedings of the 1st European conference on grain legumes. L'Association Europeenne desProteagineux, Angers, pp 417–418

Malav OP, Talukder S, Gokulakrishnan P, Chand S (2015) Meat analog: a review. Crit Rev Food Sci Nutr 55:1241–1245

Maurer D (2002) Vegetarianism: movement or moment? Temple Univ Press, Philadelphia

McMichael AJ, Powles JW, Butler CD, Uauy R (2007) Food, livestock production, energy, climate change, and health. Lancet 370:1253–1263

Muguerza E, Gimeno O, Ansorena D, Astiasaran I (2004) New formulations for healthier dry fermented sausages: a review. Trends Food Sci Technol 15:452–457

Myers W (2002) Sustainable food security for all by 2020. Proceedings of an international conference. IFPRI, Washington, DC, p 100

Nordic Council of Ministers (2014) Nordic nutrition recommendations 2012—integrating nutrition and physical activity. Copenhagen 5(11):1

Nout MJR, Aidoo KE (2002) Asian fungal fermented food. In: Indus-trial applications. Springer, New York, pp 23–47

Oboh G, Ademiluyi AO, Akindahunsi AA (2009) Changes in polyphenols distribution and antioxidant activity during fermentation of some underutilized legumes. Food Sci Technol Int 15:41–46

Okubo K, Iijima M, Kobayashi Y, YoshikoshiI M, Uchida T, Kubo S (1992) Components responsible for the undesirable taste of soybean seeds. Biosci Biotech Bioch 56:99–103

Paddon-Jones D, Westman E, Mattes RD, Wolfe RR, Astrup A, Westerterp-Plantenga M (2008) Protein, weight management, and satiety. Am J Clin Nutr 87:1558S–1561S

Pan A, Sun Q, Bernstein AM, Schulze MB, Manson JE, Stampfer MJ, Willett WC, Hu FB (2012) Red meat consumption and mortality: results from 2 prospective cohort studies. Arch Intern Med 172:555–563

Piazza J, Ruby MB, Loughnan S, Luong M, Kulik J, Watkins HM, Seigerman M (2015) Rationalizing meat consumption. The 4Ns. Appetite 91:114–128

Pimentel D, Pimentel M (2003) Sustainability of meat-based and plant-based diets and the environment. Am J Clin Nutr 78(Suppl):60S–663S

Pitesky ME, Stackhouse KR, Mitloehner FM (2009) Chapter 1-clearing the air: Livestock's contribution to climate change. In: Donald LS (ed) Advances in agronomy. Elsevier, Burlington, pp 1–40

Raats J (2007) Meat (substitutes) comparing environmental impacts. A Case study comparing Quorn and pork. University of Groningen, Groningen

Riascos JJ, Weissinger AK, Weissinger SM, Burks AW (2010) Hypoallergenic legume crops and food allergy: factors affecting feasibility and risk. J Agric Food Chem 58:20–27

Riaz MN (2004) Texturized soy protein as an ingredient. In: Yada RY (ed) Proteins in food processing. Woodhead Publishing Limited, Cambridge, pp 517–557

Ritchie H, Laird J, Ritchie D (2017) 3f bio: Halving the cost of myco-protein through integrated fermentation processes. Ind Biotechnol 13:29–31

Rodger G (2001) Production and properties of mycoprotein as a meat alternative. Food Technol Chicago 55:36–41

Rohrmann S, Overvad K, Bueno-de-Mesquita HB et al (2013) Meat consumption and mortality—results from the European prospective investigation into cancer and nutrition. BMC Med 11:63

Röös E, Sundberg C, Tidåker P, Strid I, Hansson PA (2013) Can carbon footprint serve as an indicator of the environmental impact of meat production? Ecol Indic 24:573–581

Roth DM, Mckeith FK, Brewer MS (1997) Processing parameter effects on textural characteristics of reduced-fat pork sausage. J Food Quality 20:567–574

Ruxton CHS, McMillan B (2010) The impact of mycoprotein on blood cholesterol levels: a pilot study. Br Food J 112:1092–1101

Sadler MJ (2004) Meat alternatives – market developments and health benefits. Trends Food Sci Technol 15(5):250–260

Satari B, Karimi K (2018) Mucoralean fungi for sustainable production of bioethanol and biologically active molecules. Appl Microbiol Biotechnol 102:1097–1117

Singh P, Kumar R, Sabapathy SN, Bawa AS (2008) Functional and edible uses of soy protein products. CRFSFS 7:14–28

Slavin J, Green H (2007) Dietary fibre and satiety. Nutr Bull 32:32–42

Smetana S, Mathys A, Knoch A, Heinz V (2015) Meat alternatives: life cycle assessment of most known meat substitutes. Int J Life Cycle Assess 20:1254–1267

Smetana S, Aganovic K, Irmscher S, Heinz V (2018) Agri-food waste streams utilization for development of more sustainable food substitutes. Designing sustainable technologies, products and policies: from science to innovation. Springer, Cham, pp 145–155

Smith H, Doyle S, Murphy R (2015) Filamentous fungi as a source of natural antioxidants. Food Chem 185:389–397

Suberbie F, Mendizabal D, Mendizabal C (1981) Germination of soybeans and its modifying effects on the quality of full-fat soy flour. J Am Oil Chem Soc 58:192–195

Tscharntke T, Clough Y, Wanger TC, Jackson L, Motzke I, Perfecto I, Vandermeer J, Whitbread A (2012) Global food security, biodiversity conservation and the future of agricultural intensification. Biol Conserv 151:53–59

Tuomisto H, Hanna LT, Mattos MJTD (2011) Environmental impacts of cultured meat production. Environ Sci Technol 45:6117–6121

Turnbull WH, Leeds AR, Edwards GD (1992) Mycoprotein reduced blood lipids in free-living subjects. Am J Clin Nutr 55:415–419

WHO (2003) Diet, nutrition and the prevention of chronic diseases. WHO Technical Report Series 916, Geneva

Wiebe M (2002) Myco-protein from Fusarium venenatum: a well established product for human consumption. Appl Microbiol Biotechnol 58:421–427

Williamson DA, Geiselman PJ, Lovejoy J, Greenway F, Volaufova J, Martin CK, Arnett C, Ortego L (2006) Effects of consuming mycoprotein, tofu or chicken upon subsequent eating behaviour, hunger and safety. Appetite 46:41–48

Xiong YL (2004) Muscle protein. In: Yada RY (ed) Proteins in food processing. Woodhead Publishing Limited, Cambridge, pp 100–122

Chapter 12
Fungal Mycotoxins

Neveen M. Abdelmotilib, Amira G. Darwish, Ahmed M. Abdel-Azeem, and Donia H. Sheir

12.1 Introduction

Mycotoxins (from "myco" fungus and toxin) are nonvolatile, relatively low-molecular-weight, fungal secondary metabolic products. Mycotoxins are a much-diversified group of toxic compounds produced by five kinds of spore-forming fungi, known to cause noxious effects to the health of humans and animals. Food security is regularly risked by mycotoxins appearing in food (Ilesanmi and Ilesanmi 2011).

The term mycotoxin was first used in the 1960s to describe the toxin associated with contaminated peanuts in animal feed and the loss of turkeys in England (turkey X disease). This mycotoxin was later identified as the *Aspergillus flavus* toxin aflatoxin B1. Bennett (1987) defined mycotoxins as "natural products produced by fungi that evoke a toxic response when introduced in low concentrations to higher vertebrates and other animals by a natural route." Mycotoxins are secondary metabolites, defined by Bennett and Bentley (1989) as "metabolic intermediates or products, found as a differentiation product in restricted taxonomic groups, not essential to growth and life of the producing organism, and biosynthesized from one or more general metabolites by a wider variety of pathways than is available in general

N. M. Abdelmotilib · A. G. Darwish
Food Technology Department, Arid Lands Cultivation Research Institute, City of Scientific Research and Technological Applications, Alexandria, Egypt

A. M. Abdel-Azeem (✉)
Botany and Microbiology Department, Faculty of Science, University of Suez Canal, Ismailia 41522, Egypt
e-mail: ahmed_abdelazeem@science.suez.edu.eg

D. H. Sheir
Chemistry of Natural and Microbial Products Department, Pharmaceutical Industries Division, National Research Centre, Cairo, Egypt

© The Author(s), under exclusive license to Springer Nature Switzerland AG 2021
X. Dai et al. (eds.), *Fungi in Sustainable Food Production*, Fungal Biology, https://doi.org/10.1007/978-3-030-64406-2_12

metabolism." The term was later applied to other toxic fungal natural products (Bennett and Klich 2003). Traditionally, toxigenic fungi contaminating agricultural grains have been conventionally divided into two groups those invade seed crops have been described as "field" fungi (e.g., *Cladosporium, Fusarium, Alternaria* spp.), which reputedly gain access to seeds during plant development, and "storage" fungi (e.g., *Aspergillus*; *Penicillium* spp.), which proliferate during storage (Legan 2000). Currently, this division is not so strict because according to Miller (1995) four types of toxigenic fungi can be distinguished: (1) plant pathogens as *Fusarium graminearum* and *Alternaria alternata*; (2) fungi that grow and produce mycotoxins on senescent or stressed plants, e.g., *F. moniliforme* and *Aspergillus flavus*; (3) fungi that initially colonize the plant and increase the feedstock's susceptibility to contamination after harvesting, e.g., *A. flavus*; and (4) fungi that are found on the soil or decaying plant material that occur on the developing kernels in the field and later proliferate in storage if conditions permit, e.g., *P. verrucosum* and *A. ochraceus*. The involvement of *Aspergillus* spp. as plant pathogens has been reported, and aflatoxin-infected crops have from time to time been returned to agricultural soils. This practice may prove hazardous, since both *A. flavus* and *A. parasiticus* can infect crops prior to harvesting (Lillehoj and Zuber 1973). The phytotoxic effects of the aflatoxins have been investigated, with respect to seed germination and the inhibition of root and hypocotyl elongation (Llewellyn et al. 1984; McLean et al. 1992). Aflatoxin has been reported to occur within apparently healthy, intact seeds which suggest that the toxin can be transported from contaminated soil to the fruit (Anderson et al. 1975). Aflatoxin B1 (AFB1) can be translocated from the roots to the stems and leaves. If the soil microorganisms do not rapidly degrade the aflatoxin contained within the plowed under stover and grains, the possibility that the roots of the seedlings of the following year's crop will both absorb and translocate the aflatoxins to both the stems and leaves exists (Mertz et al. 1980). This could be hazardous to the plant's growth and development as well as to the consumer's health. The *Penicillium* genus dominated the fungal flora, with mycotoxigenic species such as *P. crustosum, P. chrysogenum, P. hirsutum, P. expansum, P. roqueforti, P. viridicatum, P. commune, P. aurantiogriseum, P. citrinum, P. verrucosum, P. cyclopium, P. canescens, P. madriti, P. palitans, P. thomii, P. baarnense, P. fenneliae*, and *P. frequentans*. These fungi have been reported to produce a number of toxins as citrinin (CTN), cyclopiazonic acid (CPA), ochratoxin A (OTA), patulin (PAT), penicillic acid (PA), penitrem A (PNT), roquefortine (RQF), frequentin (FRE), palitantin (PAL), mycophenolic acid (MPA), viomellein (VIM), gliotoxin (GT), citreoviridin (CIV), and rubratoxin B (RB) (Frisvad and Filtenborg 1983; Frisvad and Thrane 1996; Frisvad and Samson 2004; Yamaji et al. 2005; Ismaiel and Papenbrock 2014).

These fungal species and their mycotoxins contaminate harvested seeds causing losses of agricultural commodities in many zones of the world. Such contaminants are fearsome, since they affect the seeds before harvest time and may find optimal developing conditions when the seeds are stored, leading to alteration of the germination quality of these seeds (Koteswara Rao et al. 2014). *Alternaria, Helminthosporium, Pyrenophora* (sexual state: *Drechslera*), *Phoma*, and *Zygosporium* are genera of saprobic and plant pathogenic dematiaceous fungi with a worldwide distribution, commonly associated with leaves, wood, cereals, and other grasses. The species of these fungal genera are known to produce some dangerous mycotoxins such as cytochalasins and tenuazonic acid (TA). Members of the genus *Fusarium* produce a range of chemically different phytotoxic compounds, such as fusaric acid (FA), fumonisins (fumonisin B1, FB1), beauvericin (BEA), enniatin (ENN), moniliformin (MON), and trichothecenes. These possess a variety of biological activities and cause morphological, physiological, and metabolic effects including necrosis, chlorosis, growth inhibition, wilting, and inhibition of seed germination and effects on calli (Desjardins and Hohn 1997).

Table 12.1 summarizes the general classes of mycotoxins with their producing fungal genera/species and chemical groups. Some of these mycotoxins are phytotoxic and others are non-phytotoxic. Mycotoxins with phytotoxic properties will be addressed below on the basis of their producing species. This review provides insight into characterization and evaluation of mycotoxins as phytotoxins. For each of the diverse groups of mycotoxins, we will give examples indicating history and fungi production with a focus on their modes of action and phytotoxic effects on several morphological and biochemical processes.

12.2   Types of Mycotoxins

There are five mycotoxins or groups of mycotoxins that occur quite often in food: deoxynivalenol/nivalenol, zearalenone, ochratoxin, fumonisins, and aflatoxins. The fungi that produce mycotoxins in food fall broadly into two groups: those that invade before harvest, commonly called field fungi, and those that occur only after harvest, called storage fungi. There are three types of toxicogenic field fungi: plant pathogens such as *Fusarium graminearum* (deoxynivalenol, nivalenol); fungi that grow on senescent or stressed plants, such as *Fusarium moniliforme* (fumonisin) and sometimes *Aspergillus flavus* (aflatoxin); and fungi that initially colonize the plant before harvest and predispose the commodity to mycotoxin contamination after harvest, such as *Penicillium verrucosum* (ochratoxin) and *A. flavus* (aflatoxin) (Tola and Kebede 2016).

12.3   Occurrence and Toxicity of Mycotoxins

Mycotoxins are poisonous chemical compounds and secondary metabolites produced by fungus or molds. Those mycotoxins that do occur in food and/or feedstuffs have great significance in the health of humans and livestock. Since they are produced by fungi, mycotoxins are associated with diseased or moldy crops, although the visible mold contamination can be superficial. The effects of some foodborne mycotoxins are acute symptoms of severe illness appearing very quickly. Other

Table 12.1 Common types of mycotoxins, taxa producing them, and their chemical groups

Name	Species/genus/group	Chemical group
Citreoviridin	*Penicillium citreo-viride*	Pyranone derivative
Citrinin	*Penicillium, Monascus, Aspergillus terreus*	Benzopyran compound
Cyclopiazonic acid	*Penicillium, Aspergillus*	Indole tetramic acid
Frequentin	*P. frequentans*	Carbocyclic compound
Gliotoxin	*A. fumigatus*	Epipolythiodioxopiperazine
Mycophenolic acid	*Penicillium*	Meroterpenoid compound
Ochratoxin A	*Aspergillus, Penicillium*	Benzopyran compound
Palitantin	*Penicillium*	Cyclohexane derivative
Patulin	*Penicillium, Aspergillus*	Benzopyran compound
Penicillic acid	*Penicillium, Aspergillus*	Isopropylidene tetronic acid
Penitrem A	*Penicillium*	Indole diterpene alkaloid
Roquefortine	*Penicillium*	Diketopiperazine compound
Rubratoxin B	*Penicillium rubrum*	Alpha, beta unsaturated lactone
Viomellein	*Penicillium*	Benzopyran compound
Acetyldeoxynivalenol (Adeyeye 2016)	*Fusarium*	Trichothecenes, sesquiterpenoid
Deoxynivalenol (Ahlberg et al. 2015)	*Fusarium*	Trichothecenes, sesquiterpenoid
Beauvericin	*Fusarium*	Hexadepsipeptide compound
Diacetoxyscirpenol	*Fusarium*	Trichothecenes, sesquiterpenoid
Enniatin	*Fusarium*	Cyclic depsipeptide
Fumonisins	*Fusarium*	Monoterpenes
Fusaric acid	*Fusarium*	Picolinic acid derivative, carboxylic acid
HT-2 toxin	*Fusarium*	Trichothecenes, sesquiterpenoid
Moniliformin	*Fusarium*	Cyclobutane compound, dione
Nivalenol	*Fusarium*	Trichothecenes, sesquiterpenoid
T-2 toxin	*Fusarium*	Trichothecenes, sesquiterpenoid
Zearalenone/F-2 toxin	*Fusarium*	Estrogenic compound
Aflatoxin	*Aspergillus*	Difuranocoumarin derivative
Tenuazonic acid	*Alternaria tenuis*	Acetyltetramic acid (Adeyeye 2016)
Cytochalasin	*Phoma*	Polyketide-amino acid hybrid

mycotoxins occurring in food have longer-term chronic or cumulative effects on health, including the induction of cancers and immune deficiency (Ayalew 2010).

Many of these mycotoxins can cause serious problems in livestock resulting in substantial economic losses. The most common mycotoxins are aflatoxins, ochratoxins, trichothecenes, fumonisins, zearalenone, and ergot alkaloids. These mycotoxins are carcinogenic, mutagenic, teratogenic, and immunosuppressive (Es'haghi et al. 2011).

Mycotoxins are harmful substances produced by fungi in various foods and are estimated to affect as much as 25% of the world's crop each year. Most of these mycotoxins belong to the three genera of fungi: *Aspergillus*, *Penicillium*, and *Fusarium*. Although over 300 mycotoxins are known, those of most concern based on their toxicity and occurrence are aflatoxin, vomitoxin, ochratoxin, zearalenone, fumonisin, and T-2 toxin. They are produced in cereal grains as well as forages before, during, and after harvest in various environmental conditions. The presence of mycotoxins in feeds may decrease feed intake and affect animal performance. In addition, the possible presence of toxic residues in edible animal products such as milk, meat, and eggs may have some detrimental effects on human health. Fungal contamination affects both the organoleptic characteristics and the alimentary value of feeds and entails a risk of toxicosis. The biological effects of mycotoxin depend on the ingested amounts, number of occurring toxins, duration of exposure to myco-toxin, and animal sensitivity. Mycotoxins display a diversity of chemical structures, accounting for their different biological effects. Depending on their precise nature, these toxins may be carcinogenic, teratogenic, mutagenic, immunosuppressive, tremor genic, hemorrhagic, hepatotoxic, nephrotoxic, and neurotoxic. Mycotoxins are metabolized in the liver and the kidneys and also by microorganisms in the digestive tract. Therefore, often the chemical structure and associated toxicity of mycotoxin residues excreted by animals or found in their tissues are different from the parent molecule (Akande et al. 2006).

12.4 The Factors Affecting Mycotoxin Production

Mycotoxins to human and animal health have multiple factors affecting production and/or presence of mycotoxins in foods or feeds. Hence, isolation and confirmation of mycotoxigenic fungal species in foods or feeds don't indicate the presence of mycotoxins. Upon development of accurate and sensitive techniques for qualitative and quantitative analysis of mycotoxins, researchers have found that various factors operate interdependently to affect fungal colonization and/or production of the mycotoxins. Factors that affect mycotoxin production and contamination can be categorized as physical, chemical, and biological. Physical factors include environmental conditions conducive to fungal colonization and mycotoxin production such as temperature, relative humidity, and insect infestation. Chemical factors include the use of fungicides and/or fertilizers. Biological factors are based on the interaction between the colonizing toxigenic fungal species and substrate (Tola and Kebede 2016).

Slightly higher CO_2 concentrations, interactions with the temperature, and the availability of water can stimulate the growth of some mycotoxigenic species, especially under hydrous stress (Magan et al. 2011).

12.5   Control of Mycotoxins

Control of mycotoxins is for the purpose of public health importance and economic improvement in the country. Hence, a number of strategies for reduction and control of mycotoxins have been considered in different areas of the world including African countries. The control of mycotoxins in Africa involves (1) prevention of mold or fungus growth in crops and other feedstuffs; (2) decontamination of mycotoxin-contaminated feeds/foods as a secondary strategy; and (3) continuous surveillance of mycotoxins in agricultural crops, animal feedstuffs, and human foods (Tola and Kebede 2016).

Control of mold growth in foods and feeds can be accomplished by keeping moisture low, feed fresh, and equipment clean and using mold inhibitors. In addition, control of mycotoxins in animal diets would reduce the likelihood that mycotoxin residues would appear in animal products destined for human consumption (Akande et al. 2006).

Several preventive measures to minimize mycotoxin contamination in agricultural commodities have been attempted. These can be divided into three broad categories:

- Plant breeding
- Good agronomic practices
- Detoxification

The problem of ergot contamination of cereals and millets has been successfully minimized in the past by cultivating varieties of rye, wheat, and pearl millet that are resistant to the disease. The periodic re-emergence of the problem can be attributed to the release of varieties which are not resistant to ergot. However, there has been little success in providing resistant varieties of corn and peanut to minimize the problem of aflatoxins. Other agronomic approaches such as avoiding water stress, minimizing insect infestation, and reducing inoculum potential have been suggested and are effective when farmers can implement such practices. Following good agricultural practices during both preharvest and postharvest conditions would minimize the problem of contamination by mycotoxins such as aflatoxins, ochratoxin, and trichothecene mycotoxins. These include appropriate drying techniques, maintaining proper storage facilities, and taking care not to expose the grains or oilseeds to moisture during transport and marketing. The method of segregating contaminated, moldy, shriveled, or insect-infested seeds from sound kernels has been particularly useful in minimizing aflatoxin contamination in peanuts (Adeyeye 2016; Fox and Howlett 2008).

12.6   Removal of Mycotoxins in Foods

In the feed and food industry, it has become common practice to add mycotoxin binding agents such as montmorillonite or bentonite clay in order to affectively adsorb the mycotoxins. To reverse the adverse effects of mycotoxins, the following criteria are used to evaluate the functionality of any binding additive:

- Efficacy of active component verified by scientific data
- A low effective inclusion rate
- Stability over a wide pH range
- High capacity to absorb high concentrations of mycotoxins
- High affinity to absorb low concentrations of mycotoxins
- Affirmation of chemical interaction between mycotoxin and adsorbent
- Proven in vivo data with all major mycotoxins
- Nontoxic, environmentally friendly component

Since not all mycotoxins can be bound to such agents, the latest approach to mycotoxin control is mycotoxin deactivation. By means of enzymes (esterase, de-epoxidase), yeast (*Trichosporon mycotoxinvorans*), or bacterial strains (Eubacterium BBSH 797), mycotoxins can be reduced during preharvesting contamination. Other removal methods include physical separation, washing, milling, nixtamalization, heat treatment, radiation, extraction with solvents, and the use of chemical or biological agents. Irradiation methods have proven to be an effective treatment against mold growth and toxin production (Adeyeye 2016; Ashiq 2015; Kabak et al. 2006).

12.7 Mycotoxin Analysis

These methods are usually based on labor-intensive sample preparation protocols followed by traditional chromatographic separation (mostly, LC). Gas chromatography (GC) either with electron capture detection (ECD) or mass spectrometric (MS) detection is used in mycotoxin analysis, e.g., for trichothecene or battalion determination, but less frequently than alternative methods. In some cases, fast and accurate screening methods based on enzyme-linked immunosorbent assay (ELISA) are applied instead of the more labor-intensive LC methods. Thin-layer chromatography (TLC) provides a cheaper alternative to LC-based methods and has an important role, especially in developing countries, for surveillance purposes and control of regulatory limit. Modern sample cleanup techniques, such as immunoaffinity columns (IAC) or solid-phase extraction (SPE) methods, help to simplify protocols and improve selectivity and, thus, performance characteristics (Krska et al. 2008).

The introduction of ultrahigh-pressure liquid chromatography (UHPLC) has allowed faster efficient chromatographic separations, reducing runtimes. Besides, narrower peaks are obtained, which result in increased sensitivity and improved peak resolution. Some applications of this technique on mycotoxin analysis field have been recently reported in combination with tandem MS. UHPLC coupled to triple quadrupole mass spectrometer (QqQ), in selected reaction monitoring (SRM) mode, is at present considered as one of the most selective and sensitive techniques for quantification and confirmation of organic contaminants and residues in food, mycotoxin analysis included, as illustrated by the most recent scientific literature. In a recent work, we developed a rapid method with little sample manipulation for the simultaneous determination of 11 regulated mycotoxins in different food commodities by UHPLC-MS/MS. However, some difficulties had been found in order to reach the sensitivity required for aflatoxin B1 and its metabolite aflatoxin M1 in baby food, accordingly to the low concentration levels permitted by the EC (Beltrán et al. 2011).

12.8   Factors Affecting the Promulgation of Mycotoxin Regulations

A variety of factors may affect the promulgation of mycotoxin limits and regulations. These include:

- The availability of toxicological data of mycotoxins
- The availability of exposure data of mycotoxins
- Knowledge of the distribution of mycotoxin concentrations within commodity or product lots
- The availability of analytical methods
- Legislation in other countries with which trade contacts exist
- The need for sufficient food supply

The first two factors provide the information necessary for hazard assessment and exposure assessment, respectively, the main bases of risk assessment. Risk assessment is the scientific evaluation of the probability of occurrence of known or potential adverse health effects resulting from human exposure to foodborne hazards. It is the primary scientific basis for promulgation of regulations. The third and fourth factors are important factors enabling practical enforcement of mycotoxin regulations, through adequate sampling and analysis procedures. The last two factors are merely socioeconomic in nature but are equally important in the decision-making process to establish meaningful regulations and limits for mycotoxins in food and feed (Van Egmond et al. 2007).

12.9   Aflatoxins

Aflatoxins are toxic compounds produced by several species of *Aspergillus* molds. These fungi can grow under diverse conditions, but the rate of growth and aflatoxin production depends upon the prevailing physical, biological, biochemical, and environmental conditions.

Aflatoxins are secondary metabolites produced by the fungi *A. flavus* and *A. parasiticus*. The word "aflatoxin" was derived from "a" from the genus *Aspergillus*, the "fla" from the species *flavus*, and "toxin" meaning toxic. Several aflatoxins have been isolated and identified. Cole and Cox listed 16 compounds in the aflatoxin group, but only the aflatoxins B1, B2, G1, G2, and MI are routinely monitored in foods and feeds in commerce.

Monitoring foods for the presence of aflatoxins is important to ensure consumer safety. During the past two decades, several chromatographic methods have been developed for the quantification of aflatoxins in agricultural and food products. Traditionally aflatoxin analysis has been performed using solvent extraction, followed by sample cleanup by solid-phase extraction (SPE) for purification prior to analysis. The immunoaffinity technique enables a wide variety of food matrices to be analyzed using a one-step extraction protocol without the need to use halogenated hydrocarbon solvents for extraction (Otta et al. 2000).

12.9.1 Definition of Aflatoxins

Aflatoxins, difuranocoumarin compounds, are toxic secondary metabolites produced by *Aspergillus flavus* and *Aspergillus parasiticus*. They were isolated and characterized after the death of more than 100,000 turkey poults (turkey X disease) was traced to the consumption of mold-contaminated peanut meal. Thereafter, aflatoxins were found to contaminate a wide variety of agricultural products, including corn, rice, wheat, spices, and nuts (Nonaka et al. 2009).

12.9.2 Factors Promoting Fungal Growth and Aflatoxin Production

These factors could be extrinsic (temperature; relative humidity; soil properties; mechanic injury on food commodity; insect and rodent attack) or intrinsic (pH, nutrient composition, moisture content/water activity). These factors, however, do not work in isolation. Therefore, two or more factors may have to be met before fungal growth and corresponding toxin production can be effected.

12.9.2.1 Temperature

It has been reported that whether there is high or low temperature, fungal growth and its resultant mycotoxin production are inevitable. Atanda et al. observed that temperatures below 20 °C favored *Penicillium* and *Cladosporium* whereas above 20 °C enhanced growth of *Aspergillus* species. The researchers also reported that food products such as cereals and legumes were more prone to *Aspergillus* species than any other toxin-producing fungi, more so at storage due to the temperatures involved. However, fungal activity and toxin production have been reported elsewhere to be optimum at (Holcomb et al. 1992) 37 °C in the presence of other favoring conditions (Smith et al. 2016).

12.9.2.2 Moisture Content

Water content is an important factor that affects both the grade and storability of grains and legumes as it significantly influences microbial growth and toxin production. It is, thus, a key determinant of aflatoxin development in food crops. Storage fungi like *Aspergillus* require about moisture of 13% or relative humidity of 65% (water activity, a_w, of 0.65) for growth and toxin production. However, 77% or above is optimum for growth and proliferation. The researchers realized a significant (P¼ 0.000) positive correlation between *A. flavus* population and aflatoxin production, *A. flavus* population and water activity, and aflatoxin production and water activity with respective correlation coefficients of 0.849, 0.75, and 0.68. Water activity is however shown to increase with storage time; this coupled with improper drying predisposes stored cereals and legumes to fungal infestation, growth, and aflatoxin development (Okello et al. 2010).

12.9.2.3 Effect of Soil Properties on Aflatoxin Contamination in Food

Soil is another factor that has a key influence on fungal contamination in agricultural produce. Thus, crops cultivated in different soil types may have significantly varying levels of aflatoxin prevalence. According to the Codex Alimentarius Commission, light sandy soils accelerate growth of the fungi in peanuts, particularly under dry conditions, whereas heavier soils result in less contamination owing to their high water retention capacity that helps in the reduction of drought stress. Though Ghana is made of different soil types such as sandy, loamy, and clayey, the specific type of soil of a particular area may depend on the part of the country it is located. The northern part of the country has mainly sandy and sandy loamy soils, while the southern part is made of soil types ranging from clayey loamy to dark loamy (Fearon 2000; Braimoh and Vlek 2004; Abubakari 2012).

12.9.2.4 Impact of Nutrient Composition of Food
on Aflatoxin Contamination

Regardless of the fact that molds have the genetic potential to produce a particular mycotoxin, the level and rate of production would partly be influenced by available nutrients. As such, different food substrates may have different effects on aflatoxin production due to the difference in nutrient content. Findings on the effect of nutrients in substrates (corn, wheat, peanut, soybean, corn germ, and corn endosperm) on aflatoxin B1 production showed a slight *A. flavus* contamination and a relatively low level of AFB1 in defatted substrates. However, AFB1 levels sharply increased with the addition of corn oil. The levels of AFB1 in full-fat substrates were also higher than in the defatted substrates. Therefore processing complementary foods from full-fat cereals and legumes may increase the potential of aflatoxin contamination (Achaglinkame et al. 2017).

12.9.3 Classification of Aflatoxin

Among the mycotoxins, aflatoxins are most intensively sought because of their immunotoxicity acting on phagocytes and cell-mediated immunity (Bondy & Pestka 2000). Among 18 different types of the identified AFs, the major ones are AFB1 and AFB2. Due to hazardous nature of AFs to humans and animals, the European Union in the Commission Regulation No. 1881/2006 has set the maximum levels (MLs) of AFs in cereals and their derivatives as 2 g kg^{-1} for AFB1 and 4 g kg^{-1} for total AFs (AFs B1, B2, G1, and G2) (Hashemi et al. 2014). Aflatoxin M1 (AFM1) is the major oxidized metabolite of AFB1, and milk and dairy products may be contaminated by feed carry-over contamination (Campone et al. 2016).

Aflatoxins are classified according to their physical–chemical and toxicological characters in the most dangerous row of the mycotoxins. These aflatoxins are in part responsible of irreversible medical disasters that are not easily manageable such as cancer of the liver and kidneys and, in the other part, of losses in the stored cereal products (Gacem and El Hadj-Khelil 2016).

Aflatoxins (AFs) are dihydrofuranocoumarin metabolites of food fungi *Aspergillus flavus*, and *A. parasiticus*, well known for AFB1 which is most toxic. Aflatoxins B1, B2, G1, and G2 are most significant contaminants to rice. Two other metabolites, aflatoxins M1 and M2, can be separated from milk (Sun et al. 2017).

12.9.4 Structure and Function of Aflatoxin Molecules

Aflatoxins are chemically difurocoumarin derivatives derived from polyketide pathway. Recently, genetic studies by Payne and Brown (1998) on aflatoxin-producing fungi have given an interesting insight and led to the cloning of 17 genes responsible for 12 enzymatic conversions in the aflatoxin biosynthetic pathway. The physical factors influencing the pathway have been found out, but the exact role played by them is not yet clear. The pathway-specific regulation occurs by a Zn(II) 2 Cys 6 DNA binding protein that regulates the transcription of all pathway genes. Presently, 18 different types of aflatoxins have been identified, with aflatoxins B1, B2, G1, G2, M1, and M2 being the most common. Of these, aflatoxin B1 (AFB1) and G1 (AFG1) occur most frequently, with AFB1 being the most potent. Physicochemical and biochemical characteristics of the AFB1 molecule reveal two important sites for toxicological activity. The first site is the double bond in position 8,9 of the furofuran ring (Fig. 12.1). The interactions of aflatoxin, DNA, and proteins, which occur at this site, alter the normal biochemical functions of these macromolecules and lead to deleterious effects at the cellular level. The second reactive group is the lactone ring in the coumarin moiety (Lee et al. 1981). The lactone ring is easily hydrolyzed and therefore is vulnerable to degradation.

There was an increase in beta structure content of arachin and conarachin II, while an aperiodic content of conarachin was observed. Kinetic studies showed the reaction to be of pseudo-first order. The modification of lysyl residues by succinylation decreased the strength of binding with conarachins, with no significant change in arachin.

Biological effects of aflatoxin can be subdivided into its toxicity, carcinogenicity, mutagenicity, and teratogenicity. The effects are influenced by species variation, sex, age, nutritional status, and effect of other chemicals. In addition, the dose level and period of exposure of the organism to the toxin are very important. The toxicological effects of AFB1 occur after the metabolic activation of the molecule by the microsomal mixed function oxidase system. These enzymatic reactions involve metabolism and detoxification. Metabolic activation of AFB1 leads to the formation of reactive AFB1-epoxide, which can lead to toxic or detoxification pathway or both. While the epoxide exerts its effect by interacting with DNA and some enzymes to alter the p-53 gene and to inhibit the enzymatic activities, it can also conjugate with proteins and glutathione, which is then excreted from the body. Various factors affecting the kinetics of formation of adducts and detoxification greatly affect the toxicity of aflatoxin and other mycotoxins. Mutation is caused by binding of the aflatoxin molecule to DNA and the subsequent erroneous protein synthesis. Sarasin et al. (1977) revealed that a DNA repair mechanism occurs in human cells after treatment with activated AFB1. Aflatoxins, being potent protein synthesis inhibitors, impair differentiation in sensitive primordial cells and lead to a teratogenic effect on certain animals. The structure-activity relationship in toxicity and carcinogenicity of aflatoxins and analogues was studied by Shank et al. (1972).

Aflatoxins may be considered biosynthetic inhibitors both in vivo and in vitro, with large doses causing the total inhibition of biochemical processes and lower doses affecting different metabolic pathways. They inhibit O_2 uptake in whole tissues by acting on adenosine triphosphatase enzyme of electron transport chain resulting in the decreased production of ATP. Aflatoxin also reduces hepatic glycogen level, probably by inhibiting glycogenesis or depression of glucose transport to liver cells or acceleration of glycogenolysis. Aflatoxin binds strongly to DNA and

Fig. 12.1 Structure of AFB1

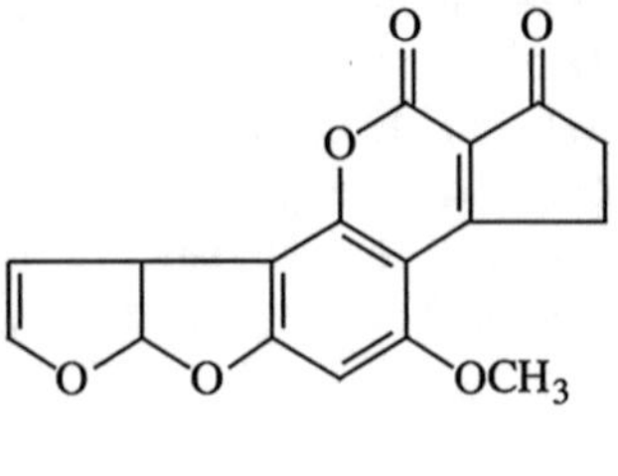

RNA. At the molecular level, aflatoxins impart their effect either interfering with DNA replication or transcription of messenger RNA into protein. Clifford et al. (1967) revealed that the administration of a single dose of aflatoxin (7 mg/kg body weight) resulted in slow development of periportal necrosis. Hepatic enzymes were released into the serum after 48 h of poisoning. This was followed by a rise in serum phosphatase activity and bilirubin concentration. Shortly after this study, it was revealed (Clifford et al. 1967) that the biochemical changes underlying the development of liver necrosis in the rat after administration of AFB1 were initiated by the toxin interacting with DNA. This interaction prevented the RNA polymerase transcribing the DNA and inhibited the formation of mRNA. The failure of mRNA formation resulted in an inhibition in protein synthesis that they considered to be the cause of the liver necrosis. They compared interactions of AFB1, AFG1, and AFG2 with DNA. Raney et al. (1993) studied the binding of AFB1 to DNA and DNA adduction by corresponding epoxide. Their results demonstrated that duplex structure favors adduct formation. Adduct yields were compared for A, B, and Z form of DNA helices. About 12 times less adduct was produced from the A-form helix when compared with the B form, while no adduct was produced from a Z-form duplex. They concluded that reaction of AFB1-8,9-epoxide with DNA proceeds via an intercalated transition state complex only with the B form of the double helix.

Aflatoxin binds with lysine component of serum albumin resulting in the formation of lysine AFB1 (Sabbioni 1990). Ch'ih et al. (1993) investigated the nuclear translocation of AFB1-protein complex. The in vitro binding of 3[H] AFB1 to various proteins was studied by equilibrium dialysis. It was reported that at 23 °C, 3[H] AFB1 binding activity (mMol/mole) decreased as follows: pyruvate kinase > albumin-NLS > albumin > carbonic anhydrase > RNase > histone. The nuclear translocation and activation of AFB1 and AFB1 protein complexes were investigated using isolated rat liver nuclei in the presence of an ATP and NADPH regenerating system (Mishra and Das 2003).

12.9.5 Occurrence of Aflatoxin

AFs may contaminate a wide variety of agricultural commodities, especially if they have high carbohydrate and/or fat contents. Cereals and all products derived from cereals represent a risk for the consumers, because these products are very sensitive to AF contamination (Anthony et al. 2012; Hashemi et al. 2014).

Aflatoxins are carcinogenic compounds, and their presence in food is a major food-related health issue. Aflatoxins are commonly found in foodstuffs such as groundnuts, wine, and maize and feed products such as wheat. Maize has the highest risk of aflatoxin contamination among cereals in the European Union, which has limited its presence to 4 l g kg⁻¹ in maize foodstuff (Verheecke et al. 2014).

AFB1 revealed that it is quite stable during heat processing operations leading to its presence in peanut products as peanut butter that is traditionally made by grinding or crushing dried and roasted groundnuts. In fact, many recent reports have revealed a co-occurrence of AFB1 in peanut products. This co-occurrence may result in some cases in synergistic action in humans and animals based on toxicological effects of AFB1 (Abel Boli et al. 2017).

Contamination by aflatoxins can take place at any point along the food chain from the field, harvest, handling, shipment, and storage. Aflatoxins have been found to contaminate a wide variety of important agricultural products worldwide, e.g., corn, wheat, rice, spices, dried fruits, and nuts. They have been clearly identified as highly toxic, mutagenic, teratogenic, and carcinogenic compounds and have been implicated as causative agents in human hepatic and extrahepatic carcinogenesis. These compounds can enter the food chain mainly by ingestion through the dietary channel of humans and animals. Different countries have imposed different legal limits on various food items and animal feeds. The level in feeds is generally higher than for human consumption (e.g., the Korean maximum legal limits for aflatoxins B1 are 10 and 50 ng g^{-1} in food and feed, respectively) (Khayoon et al. 2010).

The order of toxicity, AFB1 > AFB2 > AFG1 > AFG2, indicates that the terminal furan moiety of AFB1 is the critical point for determining the degree of biological activity of this group of mycotoxins. Aflatoxins are found generally in feed and foodstuff, such as cereal for human consumption and all products derived from cereals, including processed cereal products. It can be assumed that about 20% of food products, mainly from vegetables (e.g., cereals), are substantially contaminated. For this reason and for their toxicity, the European Union in the Commission Regulation (EC) No. 1881/2006 has established the maximum residue limits (MRLs) of aflatoxins in cereals and their derivatives: 2 g kg^{-1} for AFB1, 4 g kg^{-1} for the sum of the four aflatoxins, and 0.1 g kg^{-1} for AFB1 in processed cereal-based foods and baby foods for infants and young children (Quinto et al. 2009).

Among the four major AFs – B1, B2, G1, and G2 – aflatoxin B1 (AFB1) is the most toxic and carcinogenic. The manifestation of chronic or acute toxicosis as well as carcinogenicity depends on the dose, duration of exposure, and rate of metabolism to less toxic metabolites. Aflatoxin B1 was classified as a group 1 carcinogen (carcinogenic to humans) by the International Agency for Research on Cancer. The toxicity of AFB1 in birds has been widely investigated, being the liver the target organ for the toxin. Biochemical, hematological, immunological, and pathological effects of AFB1 have also been well-described. The liver is also the main organ where AFB1 is stored, metabolized, and/or conjugated to nucleic acids and proteins. Aflatoxin B1 is activated by cytochrome P450 enzymes (CYP), including CYP1A2, CYP3A4, and CYP2A6, and converted to epoxides (AFB1-8,(Ayalew 2010) exo-epoxide, and AFB18 (Ayalew 2010) endo-epoxide), aflatoxin M1 (AFM1), aflatoxin P1 (AFP1), aflatoxin Q1 (AFQ1), or its reduced form aflatoxicol (AFL) (Magnoli et al. 2018).

12.9.6 Major Factors Influencing Biosynthesis of AFB1

The biosynthesis of the AFB1 requires several steps, and it is perhaps affected by the intervention of several environmental factors (stress, quorum sensing, and protein signaling pathway) without forgetting the factors regulating the transcription unit. Amino acids such as tryptophan inhibit the synthesis of aflatoxin, whereas tyrosine encourages it. The presence of the lipids induces the aflatoxinogenesis. Among the organic factors affecting biosynthesis, carbon and nitrogen are the major ones. In addition, simple sugars such as glucose and fructose support this biosynthesis, whereas in the cases of sorbose and lactose, no action has been recorded. Concerning the physical factors, the optimal temperature of biosynthesis is located between 28 °C and 35 °C. Above this temperature range, biosynthesis is inhibited due to the attack of transcription genes aflR and aflS, whereas under the conditions of dryness, the production of the aflatoxins is high. Synthesis is also influenced by subcultures and changes in the morphology of producing cells. For pH, biosynthesis is high in acidic media, while it is inhibited in basic conditions, for *A. parasiticus*, the growth in water is faster with a pH ranging from 5.5 to 6.5. The secondary plant metabolites play a key role in the synthesis of aflatoxins. For example, the presence of the octanal causes a reduction of 60% of the fungic growth with a rate of increase in the production of aflatoxins of 500%. However, hydrolysable tannins considerably inhibit the biosynthesis of aflatoxins. Some antioxidants such as the phenolic compounds, ascorbic acid and caffeic acid decrease, in an important way, the aflatoxinogenesis, without any effect on the growth of the fungi (Bueno et al. 2007).

12.9.7 Degradation of Aflatoxins

12.9.7.1 Detoxification Using Lactic Acid Bacteria

Several lactic acid bacteria are able to bind AFB1 in vitro and in vivo on the surface of the organism, and two aspects were taken into consideration: binding and release of toxin. Turbic and his collaborators showed that 77–95% of AFB1 were removed by strains of *Lactobacillus rhamnosus* GG and LC-705. El Khoury and his collaborators also noted that *Lactobacillus bulgaricus* and *Streptococcus thermophilus* were effective in the reduction of aflatoxins M1. *Lactobacillus pentosus* and *Lactobacillus brevis* have the capacity to absorb and release AFB1 (Joubrane et al. 2011; Hamidi et al. 2013). In binding of AFM1 from PBS *Lb. acidophilus*, LA1 showed a binding ability of 18.3% in viable and 25.5% in heat-killed cells (Pierides et al. 2000). El Khoury et al. (2011) studied the ability of *Lb. bulgaricus* to reduce AFM1 from PBS and yogurt. Binding was 40% after 2 h PBS incubation and increased up to 87.6% after 14 h. In yogurt the AFM1 binding reached up to 60% after a 6-h yogurt incubation. Sarimehmetoğlu and Küplülü (2004) analyzed commonly used yogurt bacteria, *Lb. delbrueckii* subsp. *bulgaricus* for its binding ability of AFM1 in PBS and in milk. Binding was better in milk (27.6%) than in PBS (18.7%) after a 4-h incubation at 37 °C. Pierides et al. (2000) tested *Lb. gasseri* for its ability to remove AFM1 from liquid PBS during 15–16-h incubation at 37 °C. Heat-killed bacteria had a better AFM1 binding ability than the viable bacteria, 61.5% and 30.8%, respectively. Pierides et al. (2000) studied the abilities of *Lb. rhamnosus* GG (ATCC 53013), *Lb. rhamnosus* LC-705, and *Lb. rhamnosus* 1/3 to bind AFM1 from PBS. *Lb. rhamnosus* GG bound over 50% of the AFM1 in PBS in all tested forms (precultured, freeze dried, viable, and heat-killed). Viable *Lb. rhamnosus* LC705 bound around 45–46% and the heat-killed more than 50%. The heat-killed *Lb. rhamnosus* 1/3 strain bound 40% and the viable 18% of the added AFM1. *Lb. rhamnosus* GG and LC-705 were further tested in skim milk and in full cream milk. *Lb. rhamnosus* GG bound with limitations: viable cells bound 19% of AFM1 in skim milk and 26% in full cream milk. The heat-killed *Lb. rhamnosus* GG bound 27% of AFM1 in skim milk and 37% in full cream milk. The viable *Lb. rhamnosus* LC-705 bound over 60% of the AFM1 in skim and full cream milk when the binding share of heat-treated cells remained at around 30% (Ahlberg et al. 2015).

12.9.7.2 Detoxification Using Food Additives

Degradation of four aflatoxins (AFB1, AFB2, AFG1, and AFG2) by food additives was investigated. Pure aflatoxins were degraded by treatment with solutions of acidic food additives (hydrochloric acid:HCl and sulfuric acid:$HzS04$), alkaline food additives (sodium bicarbonate:$NaHC0_3$, sodium carbonate:$NazC0_3$, sodium hydroxide:NaOH, sodium sulfite:Na_2S0_3, and sodium hypochlorite:NaOCl), and neutral food additives (potassium metabisulfite:K_2S_2Os, sodium bisulfite:$NaHS0_3$, sodium hydrosulfite:$Na_2S_2O_4$, hydrogen peroxide:$HzOz$, sodium chlorite:$NaCIO_2$, and ammonium peroxodisulfate:$(NH4)_2SPs$). The aflatoxins were treated with these neutral food additives under several conditions, and the effects of treatment temperature, time, and concentration of food additives on aflatoxin degradation were studied. Potassium bromate ($KBr0_3$), potassium nitrate (KNO), and sodium nitrite (NaNO) had no effect on aflatoxins. Of the aflatoxin added to corn, 20% AFB remained after treatment with the solution of $NaHS0_3$ (0.5%,48 h), but all of the AFB was completely degraded by $NaCIO_2$ (0.25%, pH 4, 48 h) and (NH4)ZSPS (0.25%, 48 h) at 60 °C. Of the aflatoxins added to butter beans, less than 20 and 5% of AFB remained after boiling treatment with a 2 and 0.5% solution of NaZSp4, respectively. These findings suggested that aflatoxins can be degraded or removed by treatment with food additives during food processing (Tabata et al. 1994).

12.9.7.3 Detoxification Using Bioactive Substances of Plants

Phenolic pulp extract of *Dialium guineense* has been proved effective for sweeping and trapping of oxygenated chemical species and the prevention of lipid peroxidation, protein oxidation, and DNA fragmentation by AFB1. Another adsorbent prepared from bagasse has been proved effective for detoxification of AFB1 in the gastrointestinal tract of chicks, and no negative symptoms was reported. Aflatoxin oxidase, an enzyme of *Armillaria tabescens*, presents a detoxifying activity toward AFB1. This reaction is dependent on oxygen and hydrogen peroxide production, which may play the crucial role in detoxification of aflatoxin oxidase. Laccase is another enzyme that has proven its detoxifying affinity for AFB1. Furthermore, manganese peroxidase is an enzyme of *Pleurotus ostreatus* which may detoxify AFB1 according to the enzyme concentration and the incubation period; this detoxification can reach 90% at 1.5 IU/mL of enzyme during 48 h of incubation. The treatment with kaempferol also decreases toxic effects of AFB1.

12.9.7.4 Detoxification Using Biomolecules of Fungi

Fungal proteins are molecules of small sizes very basic and rich in cysteine such as PgAFP, NFAP, and PC-Arctin. The NFAP induces an oxidative stress in sensitive fungi causing the apoptosis, whereas PgAFP inhibits growth in some toxigenic molds. A recent study proved a reduction growth of *A. flavus* with significant changes including several proteins at concentration higher than 9.38 mg/mL of PgAFP. Cells treated by PgAFP showed a more intense oxidation.

12.9.7.5 Detoxification Using Actinomycetes

New control strategies have been developed in recent years based on the use of actinomycetes. A study conducted on Cuban soil led to isolation of 563 actinomycetes, in which 50.7% have an antifungal activity against *Candida*, *Trichophyton mentagrophytes*, *Penicillium chrysogenum*, and *Colletotrichum musae*, probably due to the production of antibiotics belonging to the families of aminoglycoside, anthracycline, or polyether ionophore and polyene macrolide antibiotic. Another study conducted by Okudoh and Wallis in 2007 revealed that actinomycete isolates from forest soil had an antimicrobial activity better than those from the riparian soil. Isolates from poultry manure, straws, chickens, and compost soil had inhibition zones varying between 20 and 30 mm against *Candida utilis* (Gacem and El Hadj-Khelil 2016).

12.9.8 Detection of Aflatoxin in Food

Aflatoxin determination is no longer a particularly difficult task using current thin-layer chromatography (TLC), high-performance liquid chromatography (HPLC), and immunochemical techniques. Aflatoxins B1, B2, G1, G2, and MI can be readily separated and detected using either normal- or reversed-phase TLC or HPLC techniques, with HPLC becoming increasingly the method of choice. The challenges inherent in aflatoxin analysis currently include sampling, subsampling, and sample extraction methods and also the analytical variation associated with the chosen analytical method (Holcomb et al. 1992).

Other methods of detection were elaborated such as immunoaffinity column immune-enzymatic and immunochemical methods. Another very recent technique by electrochemical immunosensor sensitive to AFB1 based on carbon nanotubes with simple walled, this immunosensor was based on an indirect competitive binding. The detection limit is 3.5 pg/mL. In addition, the immunosensor was successfully applied for determination of AFB1 in corn powder, which showed a good correlation with the results obtained by high-performance liquid chromatography (Luan et al. 2015; Zhang et al. 2016).

An accurate and rapid LC-ESI-MS/MS analytical method was developed and validated for the simultaneous determination of aflatoxins B1, B2, G1, and G2 in lotus seeds. The samples were firstly extracted with methanol-water solution (80:20, v/v) and then cleaned up by immunoaffinity columns. The mass spectrometer was operated in the positive ionization electrospray (ESIþ) mode using multiple reaction monitoring (MRM) for analysis of four aflatoxins. The limits of detection (LODs) of aflatoxins B1, B2, G1, and G2 were 0.007, 0.005, 0.003, and 0.005 mg kg^1 based on a signal-to-noise ratio of 3:1, respectively. The limits of quantification (LOQs) of aflatoxins B1, B2, G1, and G2 were 0.02, 0.015, 0.01, and 0.015 mg kg^1 based on a signal-to-noise ratio of 10:1, respectively. Recoveries for samples of spiked lotus seeds were all above 66% with relative standard deviation all below 15% for all compounds. Nineteen out of twenty batches of lotus seeds collected from different drugstores or markets in China were found to be contaminated with aflatoxins at different levels ranging from 0.02 to 688.4 mg kg^1 (Liu et al. 2013).

TLC was the first method used for aflatoxin determination, but was replaced in the early 1980s in developed countries owing to the technical progress in HPLC and later in ELISA and fluorimetric techniques. These modern methods offer several advantages over TLC, but the instrumental requirements also increased. Currently available TLC, HPLC, or ELISA methods have been reviewed, and it was concluded that the determination of less than 1 ng/g aflatoxin B1 (thus supporting European legislation (e.g., EC/1525/98) and most of the other legislation worldwide) is no longer an analytical challenge. A method for aflatoxins in various food matrices, using HPLC at a level of 2 ng/g aflatoxin B1, or 4 ng/g total aflatoxins, has recently been collaboratively trial-tested and is in the process of being adopted as an official method. The main progress is based on easily available powerful analytical approaches, such as the production of specific antibodies for aflatoxins, which are used for ELISA and immunoaffinity cleanup as well as improved and well-established detection systems, for example, for increasing the fluorescence of aflatoxins. Using HPLC, the latter can be achieved by post-column derivatization by bromination by various means or irradiation by UV light. The excellent performance of mycotoxins detection techniques was referred to immunoaffinity column as mentioned by (Chiodini et al. 2006).

12.10   Ochratoxins

Ochratoxin A (OTA) and ochratoxin B (OTB) (Fig. 12.2) are naturally occurring mycotoxins produced by several species of the genera *Aspergillus* and *Penicillium* like *Aspergillus ochraceus* or *Penicillium viridicatum*. Ochratoxin is known to occur in commodities such as cereals, coffee, dried fruit, and red wine. It has been shown to be hepatotoxic, nephrotoxic, teratogenic, and carcinogenic to animals and is classified as a possible human carcinogen (category 2B) by the International Agency for Research on Cancer (IARC). Moreover, ochratoxin A is suspected to be the causative agent behind Balkan endemic nephropathy (BEN), a kidney disease encountered among the population of Southern Europe. Therefore, ochratoxin contamination is a worldwide problem concerning food and feed safety, and several countries have instituted ochratoxin restriction in nuts and grains. The EC regulations have set a maximum tolerable limit for ochratoxin A at 3 g/kg for all products derived from unprocessed cereals. In view of the recognized adverse effects caused by ochratoxin and the need for regulatory control, monitoring of its level in nuts and grain samples is important to evaluate health risks due to human consumption of these products. Therefore, a sensitive, selective, and simple method to determine the presence and level of ochratoxin in nuts and grain samples is essential (Saito et al. 2012).

The natural occurrence of OTA is generally associated with starch-rich foodstuffs such as cereals and derived products, legumes, and pulses, but it is also found in many other agricultural commodities such as coffee, cocoa, nuts, dried fruits, malt, beer, wines, and other beverages (Filali et al. 2001).

Ochratoxin A (OTA) is a potent nephrotoxin which may contaminate various foods and feed products worldwide. An assessment of OTA intake showed that, for European consumers, cereals are the most important dietary source of this mycotoxin. The contribution of cereals and derivative products to OTA exposure is probably equally important in other parts of the world, according to recent reports of OTA incidence in foods. Because of the toxic potential and high incidence of OTA in cereals, one of the first maximum levels for OTA set by the European Commission Regulation was specifically for raw cereal kernels (5 μg/kg) and for all products derived from cereals (3 μg/kg). OTA is produced by some *Penicillium* and *Aspergillus*

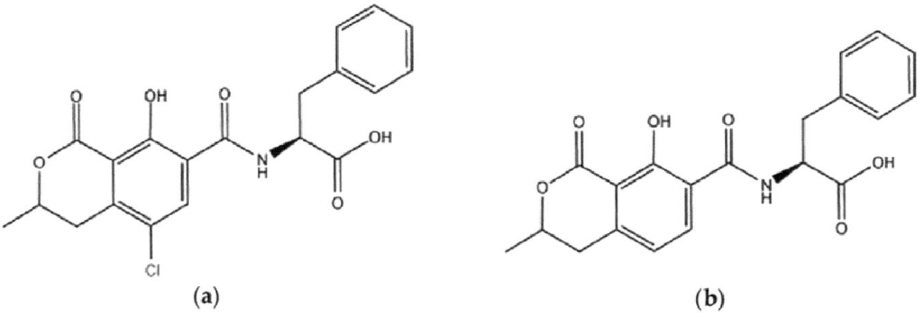

Fig. 12.2 (**a**) Structure of OTA and (**b**) structure of OTB

spp. In cool temperate zones, contamination of cereals with OTA occurs as the result of invasion by *P. verrucosum*. *Aspergillus westerdijkiae* and other related species belonging to section Circumdati are found sporadically in a wide range of stored commodities, including cereals, but are seldom the cause of substantial concentrations of OTA. However, since the first description of OTA production by *Aspergillus niger* and *Aspergillus carbonarius*, members of *Aspergillus* section Nigri have achieved greater significance because of their potential to contaminate diverse food commodities with OTA (Alborch et al. 2011).

Environmental conditions such as moisture, temperature, incubation time, and substrate type as well as other factors such as the presence of competitive flora and the integrity of the seed (Marquardt and Frohlich 1992) play an important role in the colonization by *A. ochraceus* and amount of ochratoxin produced. The most important factors influencing fungal development in stored grain ecosystems are the water availability (water activity, a_w), storage temperature, and the intergranular gas composition (Ramos et al. 1998; Pardo et al. 2005).

A significant but often unrecognized toxin burden comes from coffee (Studer-Rohr et al. 1995), beer, and juices. Instant coffee is even more critical since it was shown to contain significantly higher levels of OTA than coffee prepared from roasted beans. Black tea was found not to be contaminated by OTA, whereas 42% of children's herbal teas contained relatively high toxin concentrations of up to 10 mg/kg.

OTA in meat and meat products presents a special problem (Gareis and Scheuer 2000), as OTA carryover from feed to meat has been shown experimentally (Madsen et al. 1982). The kidney is highly contaminated followed by the liver, muscle, and fat. Considerable amounts of OTA have also been found in blood (Mortensen et al. 1983), the incidence of OTA in blood sausages from swine being 77.2%, followed by liver-type sausage (67.9%), and raw sausages (46.7%). This corresponds to maximum OTA levels of 4.6 and 3.2 ppb in blood- and liver-type sausages, respectively. It was noted that OTA contamination of meat products, e.g., beef sausages, may also arise from spices carrying the mycotoxin. In lean pork, only small amounts of the mycotoxin, at a maximum concentration of 0.14 mg/kg, were detected, in as many as 17.2% of the samples. OTA was essentially absent in meat from poultry, whereas low-level OTA was present at levels around the detection limit of 0.01 mg/kg in poultry sausage.

It was already noted that the repeated uptake of food with OTA concentrations very close to or just above the recommended limit of 5 mg OTA/kg cereal or cereal products means that the acceptable daily intake (ADI) of 5 ng OTA/kg BW per day may be rapidly achieved (Petzinger and Ziegler 2000). Elimination of OTA in humans is extremely slow, since the toxin has the longest half-life known for living mammals. Repeated, almost daily uptake of OTA, therefore, will cause low albeit toxicologically relevant toxin concentrations in blood. The delayed excretion of the toxin in man may be due to reabsorption during an enterohepatic circulation, due to reabsorption from the urine after tubular secretion, and due to extensive protein binding. Since the toxin is ingested with almost every meal, humans may not be free of toxin for very long periods. The toxin has been considered by the International

Agency for Research on Cancer to be possibly carcinogenic (group 2B) for humans (IARC 1993), meaning that steady toxin exposure must be considered as a cause for serious concern. With respect to chemical carcinogens, not only dosage but, more importantly, the time-dosage profile (c 3 t product) has to be considered as relevant for tumor development. In this respect the ubiquitous presence of OTA is a subject of a toxicological debate presently.

Liver elimination of OTA is maintained by protein carriers that shuffle the toxin from its protein-bound form in blood into the hepatocyte and subsequently secrete the toxin into bile. The uptake carrier has been identified (Kontaxi et al. 1996), but less is known about the mechanism involved in the release into bile. A carrier system is also involved in the uptake of OTA by proximal tubule cells, which secrete the toxin into urine (Tsuda et al. 1999). Such systems are biological entrance gates that determine the elimination toxicokinetics of OTA and therefore have a major impact on half-life times and selective organ exposure (Petzinger and Weidenbach 2002).

12.11 Fumonisins

Fumonisins (Fig. 12.3) are synthesized by *Fusarium verticillioides* and *Fusarium proliferatum* fungi, under favorable high-temperature and humid climates, and they are frequently found in maize and other cereal grains. In humans, fumonisins are associated with neural tube defects, as well as hepatic and esophageal cancer. Among the various types, fumonisin B1 (FB1) is the most commonly found and predominant cancer promoter for humans. FB1 contamination in maize and maize-based products has been found in several countries, and it is estimated that more than half of the maize products around the world are contaminated. The FAO/WHO reported in 2011 that 63% of maize samples and 80% of the maize-based product samples were contaminated with FB1. Rice contaminated with FB1 in Japan, Korea, and Canada has also been described. In addition to maize and rice, FB1 can contaminate sorghum, wheat, barley, soybeans, black tea, and even milk.

Initially described and characterized in 1988, these substances are produced by several species of the genus *Fusarium*, especially by *Fusarium verticillioides* (previously classified as *Fusarium moniliforme*), *Fusarium proliferatum* and *Fusarium nygamai*, besides *Alternaria alternata* f.sp. *lycopersici*. Other species, such as *Fusarium anthophilum*, *Fusarium dlamini*, *Fusarium napiforme*, *Fusarium subglutinans*, *Fusarium polyphialidicum*, and *Fusarium oxysporum*, have also been included in the group of producers of these mycotoxins. Fumonisins constitute a group which include, to date, 16 substances referred to as B1 (FB1, FB2, FB3, and FB4), A1, A2, A3, AK1, C1, C3, C4, P1, P2, P3, PH1a, and PH1b. The presence of fumonisins in corn grains has been associated with cases of esophageal cancer in inhabitants of the region of Transkei in southern Africa, in China, and in northeastern Italy. Fumonisins are also responsible for the leukoencephalomalacia in equine species and rabbits; pulmonary edema and hydrothorax in pigs; and hepatotoxic, carcinogenic, and apoptosis (programmed cell death) effects in the liver of rats.

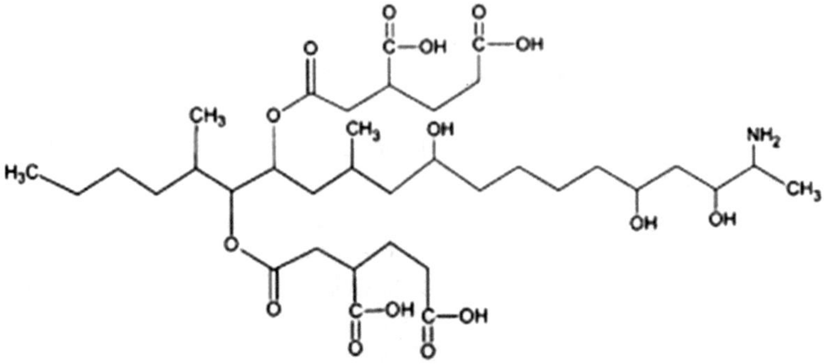

Fumonisin B₁

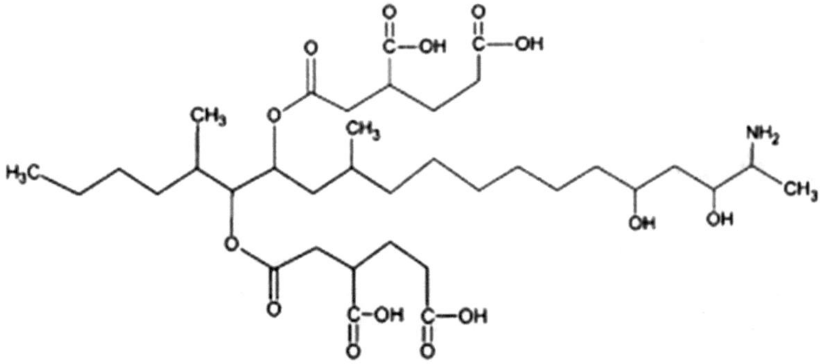

Fumonisin B₂

Fig. 12.3 Chemical structure of fumonisin B1 and B2

Fumonisins have been isolated from corn sold in a supermarket in Charleston (South Carolina), the city with the highest incidence of the occurrence of esophageal cancer among Afro-Americans in the USA. In contrast to the other mycotoxins, which are soluble in organic solvents, fumonisins are hydrosoluble, which hinders their study, and it is probable that many other mycotoxins remain undiscovered due to this hydrosolubility characteristic. The fumonisin B1, the most extensively studied, is a diester of propane 1,2,3-tricarballylic acid and (Achaglinkame et al. 2017) amino-12,16-dimethyl-3,5,10,14,15-pentahydroxycosane. The carcinogenic character of fumonisins does not seem to involve the interaction with DNA. On the other hand, its similarity with sphingosine suggests the probable intervention in the biosynthesis of sphingolipids. The inhibition of sphingolipid biosynthesis leads to serious problems related to cell activity, since these substances are essential for membrane composition, for cell-cell communication, for intracellular and cell-matrix interactions, and for growth factors. In Brazil, these mycotoxins have been detected in various substrates, particularly in corn for animal feed (da Rocha et al. 2014; Pozzi et al. 2002).

12.12   Zearalenone

Zearalenone (ZEN) (Fig. 12.4) is a nonsteroidal estrogenic mycotoxin produced by several species of *Fusarium* (*F. graminearum*, *F. crookwellemse*, *F. culmorum*, and *F. semitectum*) which primarily colonize maize and also colonize, to a lesser extent, barley, oats, wheat, and sorghum. The level of ZEN in human food can be as high as 289 g/g. Several derivatives of ZEN, including α-zearalenol (α-ZOL), and β-zearalenol (β-ZOL) as well as monohydroxylated, dihydroxylated, and formylated ZEN, have been isolated from cultures of *Fusarium*.

ZEN has predominantly estrogenic properties that are manifested in female swine, cattle, and sheep as reproductive problems. Concentrations of 1–5 g of ZEN/g of feed are sufficient to produce clinical symptoms in swine. The alpha-reduction of the keto group increases the estrogenic activity of ZEN. '-ZOL has about 10–20 times the activity of ZEN and some 100 times that of -ZOL (24). ZEN and its metabolites act as growth stimulants, and their occurrence in food has been related to the early onset of puberty in children from Puerto Rico. The co-occurrence of ZEN and trichothecenes in contaminated corn has been correlated with the incidence of human esophageal cancer in China. The estimated safe human intake of ZEN has been reported to be 0.05 g/kg of body weight/day. A small number of studies on the degradation and biotransformation of ZEN by various microorganisms have been published. El-Sharkawy et al. investigated the conversion of ZEN by 7 genera (23 species) of microorganisms. The metabolites formed included '-ZOL and -ZOL and another polar metabolite, zearalenone-(Ahlberg et al. 2015)*O*-sulfate. ZEN was reduced stereoselectively by cultures of *Candida tropicalis*, *Zygosaccharomyces rouxii*, and seven *Saccharomyces* strains to both '-ZOL and -ZOL (1). When ZEN was incubated with rumen and pig microbiota in vitro, it was also reduced to '-ZOL and –ZOL. Except for some yeast strains and some beneficial rumen microbes, none of the microorganisms tested can be used by either the food or feed industry for the purpose of ZEN detoxification. In addition, the products of the metabolism of ZEN by the tested microorganisms were more toxic or as toxic as the parent compound. We have found that specific strains of bacteria of both food and intestinal origin, and with a good safety record in the human diet, effectively bind aflatoxin and trichothecenes in vitro. In the present study, we have investigated the ability of two food-grade *Lactobacillus* strains, which are efficient in binding aflatoxins and trichothecenes, to remove ZEN and its main derivative, '-ZOL, from liquid media under variable experimental conditions (El-Nezami et al. 2002).

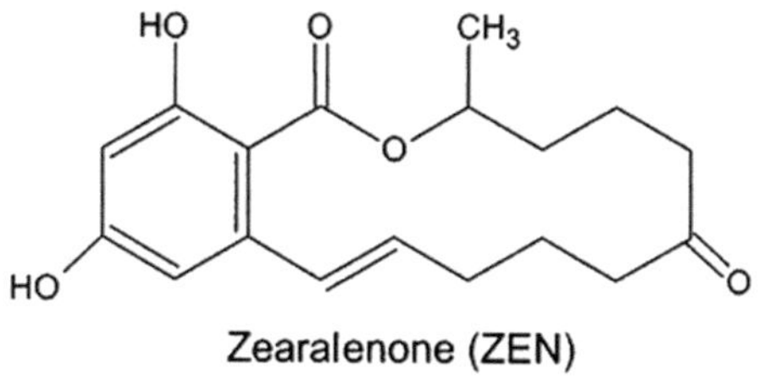

Fig. 12.4 Structure of zearalenone

Zearalenone (ZEN)

Zearalenone (ZEN) and its derivatives a- and b-zearalenol (a- and b-ZEL), zeara-lenone (ZAN), and a- and b-zearalanol (a- and b-ZAL) are mycotoxins produced by *Fusarium* species, such as *Fusarium graminearum* and *Fusarium culmorum*. They are common molds found in temperate and warm countries, and a frequent contami-nant of cereal crops, foods, and feeds worldwide posing risks to human and animal health. These nonsteroidal estrogenic mycotoxins bind competitively to estrogen receptors, and estrogenic effects and infertility caused by ZEN and its metabolites have been demonstrated in pigs, sheep, and experimental animals. In rats, both ZEN and its metabolites are transferred through the placenta. Although human exposure to ZEN has been demonstrated via dietary intake, little is known about the health effects except a few studies in cancer patients and prepubertal girls. In mammals, ZEN is metabolized to two hydroxyl isomers, a-ZEL and b-ZEL by 3a- and 3b-hydroxysteroid dehydrogenase enzymes (HSD). ZEN also undergoes minor metabolism to catechol structures. Zearalenone and its metabolites are conjugated with UDP-glucuronosyltransferases (UGTs) and sulfotransferases (SULTs). During gestation human placenta produces a considerable amount of estrogens which are essential for development of the fetus and progress of the pregnancy. Based on mycotoxins estrogenic activity human placental metabolism of ZEN and ZAN was investigated in vitro by using chorion carcinoma JEG-3 cells, human term placental subcellular fractions, and recombinant CYP19A1 enzyme (Huuskonen et al. 2015).

ZEN is an estrogenic mycotoxin mainly produced by several *Fusarium* species. ZEN-producing fungi mainly occur in wet, temperate weather and high-moisture storage environments. Exposure to ZEN has been associated with estrogenic effects, mainly reflected in precocious puberty in girls. In addition, ZEN may cause toxicity by producing reactive oxygen species (ROS). Among the cereals, maize seems to present higher infection rates by ZEN; however, wheat, soybean, and rice have been frequently found to be contaminated with ZEN. Rice, barley, and corn from Korea have been reported to be contaminated with ZEA with co-occurrence of OTA, AFB1, FB1, and DON. In Nigeria, ZEN incidence in 196 analyzed moldy rice sam-ples was 50%, with concentrations ranging from 0 to 1169 mg/kg. Milk samples were investigated in China, and the ZEN metabolite a-zearalenol, which is three times more estrogenic than ZEN, was found at levels up to 73.5 ng/kg (Luo et al. 2018).

Glucuronides are the most important class of phase II xenobiotic metabolites and as a general rule have chemical and biological properties which are significantly different from the parent compound and therefore play an important role in detoxi-fication, being particularly important for hormonal inactivation and excretion. The presence of a glucuronide moiety leads to an increased molecular mass and greater polarity. This can greatly affect a compound's ability to enter the cell and cause toxicity due to the decreased ability to diffuse across the lipids of the membrane. However, a number of glucuronides are known to be pharmacologically active, in that they retain and have enhanced or different pharmacological activities compared to the parent compound. There are two sites prone to glucuronidation present in ZEN and three in ZOL; this is due to the presence of either two or three hydroxyl groups at C-7, C-14, or C-16. The aromatic glucuronides carrying the glucuronic acid moiety at C-14 or C-16 are more polar than the aliphatic glucuronide carrying the glucuronic acid moiety at C-7. It has been demonstrated in vitro using steer, pig, rat, and human hepatic microsomes and human intestinal microsomes that ZEN and its metabolites are glucuronidated in humans and animals in the intestine and liver, preferably at the sterically unhindered (da Rocha et al. 2014) hydroxyl group resulting in, e.g., ZEN- (da Rocha et al. 2014) O-glucuronide. Human microsomes presented lower activity than those of other species, and the activity in the liver was higher than that of the intestine (Frizzell et al. 2015).

12.13   Patulin

Patulin (PAT) (Fig. 12.5) as one of these mycotoxins is an unsaturated heterocyclic lactone produced by certain fungal species of *Penicillium*, *Aspergillus*, and *Byssochlamys* growing on fruits. Patulin has been mainly found in apple and apple products and occasionally in pears, grapes, apricots, strawberries, blueberries, and peaches. As PAT is highly soluble in water and highly stable in aqueous acid media, it reaches apple derivative products, such as juices. Apple juice contaminated with PAT is hazardous for human health, not only due to the effects of PAT but also due to the toxicity produced when PAT is combined with other mycotoxins. Patulin is toxic for animals; it induces intestinal injuries, including epithelial cell degeneration, inflammation, ulceration, and hemorrhages; it has also been shown to be mutagenic, carcinogenic, immunotoxic, neurotoxic, genotoxic, and teratogenic. Acute symptoms of patulin consumption can include agitation, convulsions, edema, ulceration, and vomiting. Chronic health effects of patulin include genotoxicity, immunotoxicity, and neurotoxicity in rodents. The maximum permitted level of patulin in fruit juices and nectars, in particular apple juices and apple juice ingredients in other beverages marketed in Europe, is 50 ppb (Frizzell et al. 2015).

12.14   Conclusions

Mycotoxins have been widely studied and their implications in foods are enormous. Regulatory control and fast and effective analyses and detection will go a long way in reducing the danger of mycotoxins in foods especially in developing countries where there is shortage of food.

Fig. 12.5 Structure of patulin

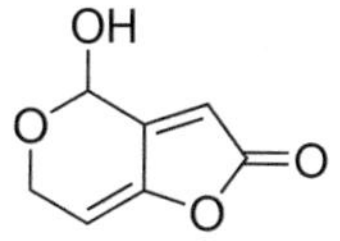

References

Abel Boli Z, Thierry Zoué L, Koussemon M, Koffi-Nevry R (2017 Jun) Low occurrence of myco-toxins in traditional peanut butter is associated with risk for consumers. Eur J Nutr Food Saf 7(4):233–243

Abubakari AH (2012) Comparative studies of soil characteristics in Shea parklands of Ghana. J Soil Sci Environ Manag 3(4)

Achaglinkame MA, Opoku N, Amagloh FK (2017) Aflatoxin contamination in cereals and legumes to reconsider usage as complementary food ingredients for Ghanaian infants: a review. J Nutr Intermed Metab 10:1–7

Adeyeye SA (2016) Fungal mycotoxins in foods: a review. Cogent Food Agric 2(1):1213127

Ahlberg SH, Joutsjoki V, Korhonen HJ (2015) Potential of lactic acid bacteria in aflatoxin risk mitigation. Int J Food Microbiol 207:87–102

Akande KE, Abubakar MM, Adegbola TA, Bogoro SE (2006) Nutritional and health implications of mycotoxins in animal feeds: a review. Pak J Nutr 5(5):398–403

Alborch L, Bragulat MR, Abarca ML, Cabañes FJ (2011) Effect of water activity, temperature and incubation time on growth and ochratoxin A production by Aspergillus niger and Aspergillus carbonarius on maize kernels. Int J Food Microbiol 147(1):53–57

Anderson HW, Nehring EW, Wichser WR (1975) Aflatoxin contamination of corn in the field. J Agric Food Chem 1975(23):775–782

Anthony MH, Francis DM, Berka NP, Ayinla GT, Haruna OG (2012) Aflatoxin contamination in foods and feeds: a special focus on Africa. In: Trends in vital food and control engineering. InTech

Ashiq S (2015) Natural occurrence of mycotoxins in food and feed: Pakistan perspective. Compr Rev Food Sci Food Saf 14(2):159–175

Ayalew A (2010) Mycotoxins and surface and internal fungi of maize from Ethiopia. Afr J Food Agric Nutr Dev 10(9):4109–4123

Beltrán E, Ibáñez M, Sancho JV, Cortés MÁ, Yusà V, Hernández F (2011) UHPLC–MS/MS highly sensitive determination of aflatoxins, the aflatoxin metabolite M1 and ochratoxin A in baby food and milk. Food Chem 126(2):737–744

Bennett JW (1987) Mycotoxins, mycotoxicoses, mycotoxicology and mycopathologia. Mycopathologia 100:3–5

Bennett JW, Bentley R (1989) What's in a name? Microbial secondary metabolism. Adv Appl Microbiol 34:1–28

Bennett JW, Klich M (2003) Mycotoxins Clin Microbiol Rev 16:497–516

Bondy GS, Pestka JJ (2000) Immunomodulation by fungal toxins. Journal of Toxicology and Environmental Health Part B. Critical review 3:109–143

Braimoh AK, Vlek PLG (2004) The impact of land-cover change on soil properties in northern Ghana. Land Degrad Dev 15(1):65e74

Bueno DJ, Casale CH, Pizzolitto RP, Salvano MA, Oliver G (2007) Physical adsorption of afla-toxin B1 by lactic acid bacteria and Saccharomyces cerevisiae: a theoretical model. J Food Prot 70(9):2148–2154

Campone L, Piccinelli AL, Celano R, Pagano I, Russo M, Rastrelli L (2016) Rapid and automated analysis of aflatoxin M1 in milk and dairy products by online solid phase extraction coupled to ultra-high-pressure-liquid-chromatography tandem mass spectrometry. J Chromatogr A 1428:212–219

Ch'ih JJ, Ewaskiewicz JI, Taggart P, Devlin TM (1993) Nuclear translocation of aflatoxin B1 – pro-tein complex. Biochem Biophsy Res Commun 190:186–191

Chiodini AM, Scherpenisse P, Bergwerff AA (2006) Ochratoxin A contents in wine: comparison of organically and conventionally produced products. J Agric Food Chem 54(19):7399–7404

Clifford JI, Rees KK, Steven MEM (1967) Effect of aflatoxins B1, G1, and G2 on protein and nucleic acid synthesis in rat liver. Biochem J 103:258

da Rocha MEB, da Chagas Oliveira Freire F, Maia FEF, Guedes MIF, Rondina D (2014) Mycotoxins and their effects on human and animal health. Food Control 36(1):159–165

Desjardins AE, Hohn TM (1997) Mycotoxins in plant pathogenesis. Mol Plant-Microbe Interact 10:147–152

El Khoury A, Atoui A, Yaghi J (2011) Analysis of aflatoxin M1 in milk and yogurt and AFM1 reduction by lactic acid bacteria used in Lebanese industry. Food Control 22(10):1695–1699

El-Nezami H, Polychronaki N, Salminen S, Mykkänen H (2002) Binding rather than metabolism may explain the interaction of two food-grade Lactobacillus strains with Zearalenone and its derivative á-Zearalenol. Appl Environ Microbiol 68(7):3545–3549

Es'haghi Z, Sorayaei H, Samadi F, Masrournia M, Bakherad Z (2011) Fabrication of a novel nano-composite based on sol–gel process for hollow fiber-solid phase microextraction of aflatoxins: B1 and B2, in cereals combined with high performance liquid chromatography–diode array detection. J Chromatogr B 879(28):3034–3040

Fearon J (2000) Economic analysis of soil conservation practices in northern Ghana. Department of Agricultural Economics, University of Ghana, p 75

Filali A, Ouammi L, Betbeder AM, Baudrimont I, Soulaymani R, Benayada A, Creppy EE (2001) Ochratoxin A in beverages from Morocco: a preliminary survey. Food Addit Contam 18(6):565–568

Fox EM, Howlett BJ (2008) Secondary metabolism: regulation and role in fungal biology. Curr Opin Microbiol 11(6):481–487

Frisvad JC, Filtenborg O (1983) Classification of terverticillate penicillia based on profiles of mycotoxins and other secondary metabolites. Appl Environ Microbiol 1983(46):1301–1310

Frisvad JC, Samson RA (2004) Polyphasic taxonomy of Penicillium subgenus Penicillium. A guide to identification of food and air-borne terverticillate Penicillia and their mycotoxins. Stud Mycol 49:1–174

Frisvad JC, Thrane U (1996) Mycotoxin production by food borne fungi. In: Samson RA, Hoekstra ES, Frisvad JC, Filtenborg O (eds) Introduction to food-borne Fungi. Centraalbureau Voor Schimmelcultures, Baarn, pp 251–260

Frizzell C, Uhlig S, Miles CO, Verhaegen S, Elliott CT, Eriksen GS, Connolly L (2015) Biotransformation of zearalenone and zearalenols to their major glucuronide metabolites reduces estrogenic activity. Toxicol In Vitro 29(3):575–581

Gacem MA, El Hadj-Khelil AO (2016) Toxicology, biosynthesis, bio-control of aflatoxin and new methods of detection. Asian Pac J Trop Biomed 6(9):808–814

Gareis M, Scheuer R (2000) Ochratoxin A in meat and meat products. Arch Leb 51:102–104

Hamidi A, Mirnejad R, Yahaghi E, Behnod V, Mirhosseini A, Amani S et al (2013) The aflatoxin B1 isolating potential of two lactic acid bacteria. Asian Pac J Trop Biomed 3(9):732–736

Hashemi M, Taherimaslak Z, Rashidi S (2014) Application of magnetic solid phase extraction for separation and determination of aflatoxins B1 and B2 in cereal products by high performance liquid chromatography-fluorescence detection. J Chromatogr B 960:200–208

Holcomb M, Wilson DM, Trucksess MW, Thompson HC Jr (1992) Determination of aflatoxins in food products by chromatography. J Chromatogr A 624(1–2):341–352

Huuskonen P, Auriola S, Pasanen M (2015) Zearalenone metabolism in human placental subcel-lular organelles, JEG-3 cells, and recombinant CYP19A1. Placenta 36(9):1052–1055

IARC (1993) Monographs on the evaluation of carcinogenic risks to humans: some naturally occurring substances: food items and constituents, heterocyclic aromatic amines and mycotox-ins, vol 56. International Agency for Research on Cancer, Lyon, pp 1–599

Ilesanmi FF, Ilesanmi OS (2011) Knowledge of aflatoxin contamination in groundnut and the risk of its ingestion among health workers in Ibadan, Nigeria. Asian Pac J Trop Biomed 1(6):493–495

Ismaiel AA, Papenbrock J (2014) The effects of patulin from Penicillium vulpinum on seedling growth, root tip ultrastructure and glutathione content of maize. Eur J Plant Pathol 139:497–509

Joubrane K, El Khoury A, Lteif R, Rizk T, Kallassy M, Hilan C, Maroun R (2011) Occurrence of aflatoxin B1 and ochratoxin A in Lebanese cultivated wheat. Mycotoxin Res 27(4):249–257. https://doi.org/10.1007/s12550-011-0101-z. Epub 2011 May 29. PMID: 23605926

Kabak B, Dobson AD, Var IIL (2006) Strategies to prevent mycotoxin contamination of food and animal feed: a review. Crit Rev Food Sci Nutr 46(8):593–619

Khayoon WS, Saad B, Salleh B, Ismail NA, Manaf NHA, Latiff AA (2010) A reversed phase high performance liquid chromatography method for the determination of fumonisins B1 and B2 in food and feed using monolithic column and positive confirmation by liquid chromatography/tandem mass spectrometry. Anal Chim Acta 679:91–97

Kontaxi M, Eckhardt B, Hagenbuch B, Stieger B, Meier PJ, Petzinger E (1996) Uptake of the mycotoxin ochratoxin A in liver cells occurs via the cloned organic anion transporting polypeptide. J Pharmacol Exp Ther 279:1507–1513

Koteswara Rao V, Girisham S, Reddy SM (2014) Influence of different species of Penicillium and their culture filtrates on seed germination and seedling growth of sorghum. J Biochem Technol 5:832–837

Krska R, Schubert-Ullrich P, Molinelli A, Sulyok M, MacDonald S, Crews C (2008) Mycotoxin analysis: an update. Food Addit Contam 25(2):152–163

Lee L, Dunn J, DeLucca A, Ciegler A (1981) Role of lactone ring of aflatoxin B 1 in toxicity and mutagenicity. Experientia 37(1):16–17

Legan JD (2000) Cereals and cereal products. In: Lund BM, Baird-Parker TC, Gould GW (eds) The microbiological safety and quality of food. Aspen Publishers Inc., Gaithersburg, pp 759–783

Lillehoj EB, Zuber MS (1973) Aflatoxin problem in corn and possible solutions. In: Proceedings of the 30th annual corn and sorghum research conference, Chicago, IL, USA, pp 230–250. Available online: http://naldc.nal.usda.gov/naldc/download.xhtml?id=29471&content=PDF. Accessed 21 Dec 2018

Liu S, Qiu F, Kong W, Wei J, Xiao X, Yang M (2013) Development and validation of an accurate and rapid LC-ESI-MS/MS method for the simultaneous quantification of aflatoxin B1, B2, G1 and G2 in lotus seeds. Food Control 29(1):156–161

Llewellyn GC, O'Donnell WA, Dashek WV (1984) Aflatoxin influences on seed germination and root elongation by two cultivars of Glycine max and uptake of 65Zn-ZnCl2 by the cultivars. Mycopathologia 86:129–136

Luan Y, Chen Z, Xie G, Chen J, Lu A, Li C et al (2015) Rapid visual detection of aflatoxin b1 by label-free aptasensor using unmodified gold nanoparticles. J Nanosci Nanotechnol 15(2):1357–1361

Luo Y, Liu X, Li J (2018) Updating techniques on controlling mycotoxins – a review. Food Control 89:123–132

Madsen A, Hald B, Lillehoj E, Mortensen HP (1982) Feeding experiments with ochratoxin A contaminated barley for bacon pigs. 2. Naturally contaminated barley given for 6 weeks from 20 kg compared with normal barley supplemented with crystalline ochratoxin A and/or citrinin. Acta Agric Scand 32:369–372

Magan N, Medina A, Aldred D (2011) Possible climate-change effects on mycotoxin contamination of food crops pre-and postharvest. Plant Pathol 60(1):150–163

Magnoli AP, Pereyra MLG, Monge MP, Cavaglieri LR, Chiacchiera SM (2018) Validation of a liquid chromatography/tandem mass spectrometry method for the detection of aflatoxin B1 residues in broiler liver. Revista Argentina de microbiologia 50(2):157–164

Marquardt RR, Frohlich AA (1992) A review of recent advances in understanding ochratoxicosis. J Anim Sci 70:3968–3988

McLean M, Berjak P, Watt MP, Dutton MF (1992) The effects of aflatoxin B1 on immature germinating maize (Zea mays) embryos. Mycopathologia 119:181–190

Mertz D, Lee D, Zuber M, Lillehoj E (1980) Uptake and metabolism of aflatoxin by Zea mays. J Agric Food Chem 28:963–966

Miller JD (1995) Fungi and mycotoxins in grain: implications for stored product research. Journal of Stored Product Research 31:1–6

Mishra HN, Das C (2003) A review on biological control and metabolism of aflatoxin. 43:245–264

Mortensen HP, Hald B, Madsen A (1983) Feeding experiments with ochratoxin A contaminated barley for bacon pigs 5. Ochratoxin A in pig blood. Acta Agric Scand 33:235–239

Nonaka Y, Saito K, Hanioka N, Narimatsu S, Kataoka H (2009) Determination of aflatoxins in food samples by automated on-line in-tube solid-phase microextraction coupled with liquid chromatography–mass. J Chromatogr A 1216:4416–4422

Okello DK, Kaaya AN, Bisikwa J, Were M, Oloka HK (2010) Management of aflatoxins in groundnuts: a manual for farmers, processors, traders and consumers in Uganda. National Agricultural Research Organization, Entebbe

Otta KH, Papp E, Bagocsi B (2000) Determination of aflatoxins in food by overpressured-layer chromatography. J Chromatogr A 882:11–16

Pardo E, Marín S, Sanchis V, Ramos AJ (2005) Impact of relative humidity and temperature on visible fungal growth and OTA production of ochratoxigenic Aspergillus ochraceus isolates on grapes. Food Microbiology 22:383–389

Payne GA, Brown MP (1998) Genetics and physiology of aflatoxin biosynthesis. Annu Rev Phytopathol 36:329–362

Petzinger E, Weidenbach A (2002) Mycotoxins in the food chain: the role of ochratoxins. Livest Prod Sci 76(3):245–250

Petzinger E, Ziegler K (2000) Ochratoxin A from a toxicological perspective. 478 Journal of Veterinary Pharmacology and Therapeutics 23:91–98

Pierides M, El-Nezami H, Peltonen K, Salminen S, Ahokas J (2000) Ability of dairy strains of lactic acid bacteria to bind aflatoxin M1 in a food model. Journal of Food Protection 63:645–50

Pozzi CR, Arcaro JRP, Júnior IA, Fagundes H, Corrêa B (2002) Aspectos relacionados à ocorrência e mecanismo de ação de fumonisinas. Ciência Rural 32(5):901–907

Quinto M et al (2009) Determination of aflatoxins in cereal flours by solid-phase microextraction coupled with liquid chromatography and post-column photochemical derivatization-fluorescence detection. J Chromatogr A 1216(49):8636–8641

Ramos AJ, Labernia N, Marın S, Sanchis V, Magan N (1998) Effect of water activity and temperature on growth and ochratoxin production by three strains of Aspergillus ochraceus on a barley extract medium and on barley grains. Int J Food Microbiol 44(1–2):133–140

Raney VM, Harris TM, Stone MP (1993) DNA conformation mediates aflatoxin B1-DNA binding and the formation of guanine N7 adducts by aflatoxin B1 8,9-exo-epoxide. Chem Res Toxicol 6:64–68

Sabbioni G (1990) Chemical and physical properties of the major serum albumin adduct of aflatoxin B 1 and their implications for the quantification in biological samples. ChemBiolInteract 75:1–15

Saito K, Ikeuchi R, Kataoka H (2012) Determination of ochratoxins in nuts and grain samples by in-tube solid-phase microextraction coupled with liquid chromatography–mass spectrometry. J Chromatogr A 1220:1–6

Sarasin AR, Smith CA, Hanawalt PC (1977) Repair of DNA in human cells after treatment with activated aflatoxin B1. Cancer Res 37:1786

Sarımehmctoğlu B, Küplülu Ö (2004) Binding ability of Aflatoxin M1 to yoghurt bacteria. Veterinary Journal of Ankara University 51:195–198

Shank RC, Gordon JE, Wogan GN (1972) Dietary aflatoxin and human liver cancer (III). Field survey of rural Thai families for ingested aflatoxins. Food Cosmet Toxicol 10:71

Smith LE, Stasiewicz M, Hestrin R, Morales L, Mutiga S, Nelson RJ (2016) Examining environmental drivers of spatial variability in aflatoxin accumulation in Kenyan maize: potential utility in risk prediction models. Afr J Food Agric Nutr Dev 16(3):11086–11105

Studer-Rohr I, Dietrich DR, Schlatter J, Schlatter C (1995) The occurrence of ochratoxin a in coffee. Food Chem Toxicol 33(5):341–355

Sun S, Xie J, Peng T, Shao B, Zhu K, Sun Y et al (2017) Broad-spectrum immunoaffinity cleanup for the determination of aflatoxins B1, B2, G1, G2, M1, M2 in Ophiocordyceps sinensis and its pharmaceutical preparations by ultra performance liquid chromatography tandem mass spectrometry. J Chromatogr B 1068:112–118

Tabata S, Kamimura H, Ibe A, Hashimoto H, Tamura Y (1994) Degradation of aflatoxins by food additives. J Food Prot 57(1):42–47

Tola M, Kebede B (2016) Occurrence, importance and control of mycotoxins: a review. Cogent Food Agric 2(1):1191103

Tsuda M, Sekine T, Takeda M, Cha SH, Kanai Y, Kimura M, Endou H (1999) Transport of ochratoxin A by renal multispecific organic anion transporter 1. J Pharmacol Exp Ther 289:1301–1305

Van Egmond HP, Schothorst RC, Jonker MA (2007) Regulations relating to mycotoxins in food. Anal Bioanal Chem 1(389):147–157

Verheecke C, Liboz T, Darriet M, Sabaou N, Mathieu F (2014) In vitro interaction of actinomycetes isolates with Aspergillus flavus: impact on aflatoxins B 1 and B 2 production. Lett Appl Microbiol 58(6):597–603

Yamaji K, Fukushi Y, Hashidoko Y, Tahara S (2005) Penicillium frequentans from Picea glehnii seedling roots as a possible biological control agent against damping-off. Ecol Res 20:103–107

Zhang X, Li CR, Wang WC, Xue J, Huang YL, Yang XX et al (2016) A novel electrochemical immunosensor for highly sensitive detection of aflatoxin B1 in corn using single-walled carbon nanotubes/chitosan. Food Chem 192:197–202

Index

A
Acetyl-CoA carboxylase (ACC) enzyme, 120
Actinomycetes, 213
Aflatoxin B1 (AFB1), 198, 210
  administration, 209
  biosynthesis, 211
  characteristics, 207
  DNA fragmentation, 212
  pathological effects, 210
  structure, 208
  toxicological effects, 210
Aflatoxins (AFs), 12, 198, 204, 207–210
  biological effects, 208
  biosynthetic pathway, 207
  classification, 207
  contamination, 210
  definition, 205
  determination, 213
  factors, 205
  production, 206
  soil types, 206
  temperature, 205
  types, 207
  water content, 206
Agaricaceae family, 153
Agro-industrial residues, 191
Agro-industrial wastes, 100
Agronomic approaches, 202
Agro-wastes, 130
Alcoholic beverages, 30, 137
  bacterial species, 138
  beer, 138
  production, 137
Alpha-linoleic acid (ALA), 125

Alveolarin, 74
Angiotensin-I-converting enzyme (ACE)
  inhibitory activity
    cold water extract, *Grifola frondosa*, 72
    peptide production, 66
    Pholiota adiposa, 72
    Pleurotus pulmonarius, 71
    reishi, 72
    Saccharomyces cerevisiae, 72
    Tricholoma matsutake, 66
  in regulating blood pressure, 66
Angkak production, 52
Antibacterial
  Agaricus bisporus, 77
  copsin, 77
  diffusion disk technique, 77
  mushroom, *Clitocybe sinopica*, 77
  RNase, 78–80
Antifungal peptide, 74
  alveolarin, 74
  cordymin, 74
  eryngin, 74
  ganodermin, 74
  isarfelin, 75
  Thai common edible mushroom, 74
Antifungal protein, 74
Antimicrobial and antioxidant activities, 174
Antioxidants, 73
Antioxidants activity, 31
Arachidonic acid (ARA), 119
Armillaria mellea, 160
Artificial meat, 184
Artificial meat products, 189
Ascomycete, 143

X. Dai et al. (eds.), *Fungi in Sustainable Food Production*, Fungal Biology,
https://doi.org/10.1007/978-3-030-64406-2

Printed in the United States
by Baker & Taylor Publisher Services